Come With Me If You Want to Live

POLITICS, LITERATURE, & FILM

Series Editor: Lee Trepanier, Samford University

The Politics, Literature Film series is an interdisciplinary examination of the intersection of politics with literature and/or film. The series is receptive to works that use a variety of methodological approaches, focus on any period from antiquity to the present, and situate their analysis in national, comparative, or global contexts. Politics, Literature, & Film seeks to be truly interdisciplinary by including authors from all the social sciences and humanities, such as political science, sociology, psychology, literature, philosophy, history, religious studies, and law. The series is open to both American and non-American literature and film. By putting forth bold and innovative ideas that appeal to a broad range of interests, the series aims to enrich our conversations about literature, film, and their relationship to politics.

Advisory Board

Recent Titles

Come With Me If You Want to Live: The Future as Foretold in Classic Sci-fi Films by Michael Harris
The Dissident Politics in Václav Havel's Vanek Plays: Who Is Ferdinand Vanek Anyway? by Carol Strong
Irony, Liberalism, and Fantasy in the Films of Christopher Nolan by Jesse Russell
Magic in Early Modern England: Literature, Politics, and Supernatural Power by Andrew Moore
Philosophical Perspective on Film by Pedro Blas Gonzalez
Between Empire and Republic: America in the Colonial Canadian Imagination by Oana Godeanu-Kenworthy
Sorcerer: William Friedkin and the New Hollywood by Mark Wheeler

Come With Me If You Want to Live

The Future as Foretold in Classic Sci-Fi Films

Michael Harris

LEXINGTON BOOKS
Lanham • Boulder • New York • London

Published by Lexington Books
An imprint of The Rowman & Littlefield Publishing Group, Inc.
4501 Forbes Boulevard, Suite 200, Lanham, Maryland 20706
www.rowman.com

86-90 Paul Street, London EC2A 4NE

British Library Cataloguing in Publication Information Available

Library of Congress Cataloging-in-Publication Data

Names: Harris, Michael, 1971- author.
Title: Come with me if you want to live : the future as foretold in classic sci-fi films / Michael Harris.
Description: Lanham : Lexington Books, 2023. | Series: Politics, literature, & film | Includes bibliographical references and index. | Summary: "If our near future sometimes feels like a dystopian sci-fi movie, that's because it is. Come With Me If You Want to Live reveals how great science fiction films predicted the world we're increasingly living in, but how we can still avoid the worst of what they warned us about"-- Provided by publisher.
Identifiers: LCCN 2023031884 (print) | LCCN 2023031885 (ebook) | ISBN 9781666940138 (cloth ; alk. paper) | ISBN 9781666940145 (epub)
Subjects: LCSH: Dystopian films--History and criticism. | Future, The, in motion pictures. | LCGFT: Film criticism.
Classification: LCC PN1995.9.D97 H37 2023 (print) | LCC PN1995.9.D97 (ebook) | DDC 791.43/615--dc23/eng/20230724
LC record available at https://lccn.loc.gov/2023031884
LC ebook record available at https://lccn.loc.gov/2023031885

∞™ The paper used in this publication meets the minimum requirements of American National Standard for Information Sciences—Permanence of Paper for Printed Library Materials, ANSI/NISO Z39.48-1992.

Contents

Preface

"Now Read on . . . into the Fantastic World of the Future!"

That's how each of the three sections (Robots, Future Cities, Star Travel!) of *The Usborne Book of the Future* begin. Written by Kenneth Gatland and David Jefferis, it was published in 1979, and it was one of my favorite books when I was growing up.

Nothing dates faster than the future. And you can certainly play that game with *The Book of the Future*, subtitled *A Trip in Time to the Year 2000 and Beyond.* So no, our houses don't yet have automated servants handing out trays of drinks, and there aren't robot forest firefighters. Or everyday Space Shuttle flights, space elevators, factories in Earth orbit, or massive domed cities, let alone an Olympics on the Moon (anticipated for 2020). Many of the timescales for supposed technologies-to-come are way off, if they happen at all, although maybe one day we will inhabit cities in space (indeed, some of today's big tech billionaires seem to be counting on it).

But what I realize now is that, even though it was a book for children and some of its predictions didn't come to pass, in other respects the future *was* contained in its pages, and I had seen it all the way back then.

As *The Book of the Future* foresaw, we *do* carry minicomputers in our pockets, much more powerful than the 'risto' watches it featured, and our homes are full of networked gadgets. Robots have "increasingly take[n] over the jobs of skilled engineers in factories." Artificial intelligence is advancing into our lives. Shopping is increasingly online (*The Book of the Future* confidently predicted that TVs will be "used to order shopping via a computerized shopping center a few kilometers away"). Electric vehicles will eventually become widespread (the book predicted an "almost totally electric world"). Computerized weapons systems are shaping the future of war. "Electronic conferencing" is finally common (prompted by the pandemic, not merely because of presumed "convenience"). And some people are experimenting with sea-borne living, akin to the floating pyramid cities featured in the book.

That it got at least some of these predictions largely right isn't surprising. *The Book of the Future* was based on the research of scientists and organizations including Bell Aerospace, Boeing, and NASA, alongside Arthur C. Clarke and *Omni* magazine (co-author Kenneth Gatland was also a leading spaceflight theorist). But there were other things about the possible future that the book spent much less time discussing—with one important exception.

I was born in 1971. The generation before me, some of them at least, could still be excited about the future. As in *The Book of the Future*, they could plausibly believe they might one day become space tourists, as well as find ways to eliminate poverty, crime, war, and disease here on Earth. A generation later, as a child I still read comics about daring space adventurers, but I knew they had a decidedly retro, 1950s feel to them. They were yesterday's future, not the actual future.

This was because, as I grew up, the actual future seemed to get darker, it *was* darker: the possibility of nuclear annihilation in the Second Cold War, demonstrations, strikes, mass unemployment, increasing inequality, and the growing sense, as the punk movement put it, of 'no future.'

As we'll see, this was reflected—and predicted—in fiction as well. From the late 1960s onward, there was a new adultness in science fiction, including at the movies. Technical advances in filmmaking helped, but the real progress was in the subject matter: there was much less fantasy about aliens, and much more fatalism about humanity and its future (or possible lack of it).

Today, young people may be even more anxious about what's to come. Unfortunately, it's probably worse than most of them imagine. In their future, they could see civilization—at least in its present, somewhat predictable form—collapse, due to everything from climate change to wars over dwindling resources. In some of its most fundamental warnings, the more adult, dystopian sci-fi of the 1970s is fast turning into fact. This is partly what this book is about.

To understand what happened to the future, we have to go back to when I was young, to a point when two very different futures were still possible and the choice between them was being made. In between its easy-to-mock certainty about coming Lunar Olympics and interstellar starships, it was this critical choice that *The Book of the Future* really got right. What really strikes me looking back at (and at the time, as I remember), is its double-page spread titled "Two trips to the 21st century", and the stark contrast it presents between a "Garden city on a cared-for planet" and a "Polluted city of a dying world."

In the latter, there's overpopulation, vehicles trundle along still powered by gasoline (alternative energies were not pursued), the environment is dying (fake plastic trees line the roads), the air is a chokingly thick brown-orange (pedestrians wear gas masks), the brutalist urban infrastructure is decaying, and people are out of work and underfed. But in the former, fumeless hydrogen and electric powered vehicles glide through a refurbished, greenified city, beneath monorails and alongside walking and cycling lanes. The air is fresh spring blue, the plants and trees are vibrant green.

There were two possible futures back then, it said, one darkly dystopian, the other much brighter—not a whizzy 1950s Jetsons-style techno-utopia, but still a pretty good one.

What I know now, and we'll discuss in this book, is that these two possible futures were based on predictions set out in landmark publications like *The Limits to Growth*, published just a few years before in 1972. And we'll examine how, contrary to their fierce critics at the time, the predictions made in these publications are increasingly coming true.

We didn't heed these warnings, or those depicted in popular culture pretty much ever since. Fundamentally, this book is about *why*. The short answer is that, starting in the 1970s, an organized and well-funded campaign was mounted against recognizing, let alone doing anything serious about, the threats we faced, and it largely succeeded in persuading people that the future would effectively take care of itself.

Indeed, we'll see how it was these predictions, both in fiction and nonfiction, of a dark future that helped to propel—in some ways helped to create—a new political movement, which until the 1970s had been relatively marginalized. What's been neglected in many previous histories is how this movement's attractive promises—that there were no limits to growth, and that technology would solve the major problems we face—were crucial in bringing them, and a new type of politician, to power. And so we set off towards one of the two futures depicted in my children's book—the wrong one.

We can't say we weren't warned, including by some of the best science fiction films of the past fifty years, which we'll use here to discuss what's going to happen next. We'll also uncover the sometimes surprising yet hidden-in-plain-sight meanings of these films—meanings that are, however else some of these films might seem dated, now more relevant than ever (inevitably, there are major spoilers for the films).

Of course, the films represent a personal collection of favorites, and, I hope, a pretty good selection based on their predictive power, for their ability to foresee the future. They're also a way of telling the story presented here, of how we lost the future, or rather how it was taken from us. Indeed, at times they become part of the story; in some cases their warnings were (and still are) dismissed as alarmist, even part of a progressive 'plot' to end progress and immiserate millions. Obviously, their inclusion here is intended to indicate that they got more fundamentally right than wrong.

It's not the future their creators wanted then, or we want now (most of us, anyway). But the more we understand what's coming, the more we might be able to prepare for it.

So, read on, not into the fantastic, but definitely into the dramatic, world of the future.

PART I

"What . . . What Seems to Be the Problem?"

Chapter 1

"I've Seen Things You People Wouldn't Believe"

Why We're Stuck in a Blade Runner *Future That's Forty Years Old*

We start in *Blade Runner* (1982), because that's where today's visions of the future started—and effectively ended. *Blade Runner* persists because its future remains our most likely future, and increasingly our present. The film vividly depicts a future of urban decay and under-investment, overpopulation, corporate domination, commodification, and alienation. It's the distillation of all dystopias—the reason it still feels relevant. So the more fundamental question, and the central theme of this book, is why we don't feel we can change these things, which is really a political question—and also the outcome of a particular political project.

"LOS ANGELES, NOVEMBER, 2019"

There's the famous opening vista, described in the script as a "vast plain of industrialization, menacing shapes on the horizon, stacks belching flames five hundred feet into the sky the color of cigar ash", crisscrossed by flying personal vehicles called spinners. Then we move toward the Tyrell Corporation's towering pyramid. Inside, a Blade Runner, Holden (Morgan Paull), is interviewing a new employee using a special device, a Voight-Kampff analyzer. It's designed to detect physical responses in test subjects to questions meant to cause an emotional reaction. After a few questions, the man being tested, Leon Kowalski (Brion James), becomes agitated. Suddenly, he stands and blasts Holden backwards through the wall.

We have to start in *Blade Runner*, because that's where today's visions of the future really started—and as we'll discuss in this chapter, effectively *ended.* There's a thousand (and many more) analyses of the film, from fan sites and entertainment blogs, to highly literary and impenetrable theoretical commentaries. But the obvious fact, hidden in plain sight, is that *Blade Runner* persists because its future remains our most likely future, and in many ways our present. It tells us what it's always told us, and most importantly, points towards why we didn't listen, why we didn't try to change the future it warned us about.

In some ways, it's strange that *Blade Runner* persists. It's not the most entertaining film; it's cold, dour, and distant, with at times a faux seriousness and profundity. It lacks an engaging protagonist and has an air of fundamental fatalism. As the author of the definitive book on the movie, *Future Noir: The Making of Blade Runner* (1996), Paul M. Sammon, puts it, in terms of people's reactions it's a "love it or who cares?" kind of experience. Neither was it the first film to depict the future in dark, gritty, disturbing terms; as we'll see, in many ways *Blade Runner* represents the culmination of over a decade of pessimistically prescient sci-fi. Unsurprisingly, at the time it wasn't a commercial success.

But from its opening moments, it has amazing production design. There's a solidity and credibility to its future. As is well known, the industrial designer and 'visual futurist' Syd Mead played a critical role in this. His modernist, optimistic commercial design work (often focusing on urban transportation) was incorporated into a dank and dying world in which the future isn't what it once was, what was promised (Mead liked to call science fiction "reality ahead of schedule"). And it's this that points to the film's recurring relevance.

Blade Runner vividly depicts a future of urban decay and under-investment, overpopulation, corporate domination, commodification, and alienation. The lack of flying cars in our own time isn't the point. Described by director Ridley Scott as "Hong Kong on a very bad day," the film depicts a future Los Angles of soaring inequality, environmental degradation, pollution, urban sprawl, absent government, and (literally) dehumanizing technological progress. It's the distillation of all dystopias—the reason it still feels relevant, even if (or perhaps *because*) it might now be too late to act as a warning.

"NOT VERY SPORTING TO FIRE ON AN UNARMED OPPONENT"

Rick Deckard (Harrison Ford), a former Blade Runner, eats on the street at a noodle bar, when Gaff (Edward James Olmos) tells him he's under arrest. Deckard tries to ignore Gaff, but when he hears that Bryant is asking for him, he gives in. Gaff flies Deckard to the police headquarters. Police boss Bryant (M. Emmet Walsh) tells Deckard there are four "skin jobs" (derogatory slang for Replicants) loose, and he needs him to retire them. Deckard refuses—until Bryant pressures him into accepting with a veiled threat.

As Scott remarks, "I've said this before, but a lot of the overall environment we came up with for *Blade Runner* was simply a forecasting of what I'd already seen in high density urban centers throughout the world" – the financial center of Hong Kong, Tokyo's Ginza district, London's Piccadilly Circus, New York's Times Square, and Milan's business area.

The Philip K. Dick novel on which the film is loosely based (the script went through many different versions), *Do Androids Dream of Electric Sheep?*, was written in the late 1960s—a time of rising environmental consciousness. Alongside its themes of artificial intelligence and "future shock" (perhaps post-traumatic stress disorder is a better way of putting it), its environmental theme was there from the beginning (Dick stated that he was particularly provoked by the dumping of barrels of nuclear waste into the oceans and how this would affect future generations).

Hence, in this future, artificial animals are created as expensive playthings for the rich. And as we'll see, the major themes of Dick's 1968 novel, those of consciousness and ecology (people prove their humanity by how much they care for animals), would in the 1970s be at the center of a battle for the future, one that ultimately would be lost.

Bryant explains that six Nexus-6 Replicants murdered twenty-three people on an off-world colony, then stole a shuttle to Earth. Four Replicants survived and disappeared. Suspecting they might try to infiltrate the Tyrell Corporation, Holden was testing new employees. Bryant tells Deckard that although the Nexus-6s are designed without emotions, they're so advanced they develop them within a few years. As a safety mechanism, they're designed with only a four-year lifespan. But why they risked returning to Earth is a mystery.

Indeed, because *Blade Runner*'s Los Angeles is both overcrowded and strangely depopulated. The streets are packed with people, but old buildings are vacant and crumbling, rain pouring relentlessly through collapsing ceilings. The rich and healthy former residents have left the planet for off-world colonies; it's mainly only the poor and the broken who are left.

One detail that's often missed in this obsessed-over film is that the capsules seductively swallowed by beautifully made-up geisha girls displayed on towering video screens on the sides of skyscrapers are birth-control pills, presumably public information pleas to try to manage this already heaving metropolis. Unfortunately, the film employs a dated shorthand of using predominantly Asian crowds to signify urban overpopulation, as well as perhaps pointing to Japan's rise as a global economic power in the 1980s.

Even so, *Blade Runner* is still our vision of the future because the future it was warning us about hasn't changed—because the ideological political-economic project that came to the fore around the time the film was released has determined our future, is still with us, and still dominates.

"WAKE UP! TIME TO DIE!"

Since *Blade Runner* is set in 2019 (the seminal 1988 Japanese anime *Akira* also takes place in a dystopian 2019), it's also our past, a future that's already happened, one we can't rewrite. The Future hasn't changed, because the conceivable future hasn't changed.

This is all the more incredible when we consider both sci-fi forty years before *Blade Runner* (which, with a few notable exceptions, bears no relation to our world), and because we were warned by both fictional and nonfictional accounts about where we were heading—and now, where we've (nearly) arrived. Sci-fi dystopias are usually comfortably decades in the future. Until suddenly, they're not.

So why didn't we listen? In the 2015 movie *Tomorrowland* (named after the futuristic land in Disney theme parks), the character played by Hugh Laurie, the governor of a once fantastical city which is now in a state of decay, laments the failure of his plan to prevent a catastrophic future by projecting images of disaster to humanity as a warning. Instead, as he notes, "They didn't fear their demise, they repackaged it. It can be enjoyed as video games, as TV shows, books, movies. The entire world wholeheartedly embraced the apocalypse and sprinted towards it with gleeful abandon."

But were the dystopic visions, or the audiences that consumed them, most to blame?

To Pauline Kael's (longtime film critic for *The New Yorker*) famous review.[1] Kael's influence casts as long a shadow on film criticism as *Blade Runner* does on contemporary sci-fi. She recognized the significance of its groundbreaking production design, of course: "The congested-megalopolis sets are extraordinary, and they're lovingly, perhaps obsessively, detailed; this is the future as a black market, made up of scrambled sordid aspects of the past." But Kael dismissed its story: "[W]e're not caught up in the pulpy suspense plot . . . *Blade Runner* is a suspenseless thriller; it appears to be a victim of its own imaginative use of hardware and miniatures and mattes." Reacting to its lack of warmth, Kael's most controversial line was catty, as she could be: "If anybody comes around with a test to detect humanoids, maybe Ridley Scott and his associates should hide." But her most astute point was about the film's fatalism: "Here we are—only forty years from now—in a horrible electronic slum, and *Blade Runner* never asks, 'How did this happen?' The picture treats this grimy, retrograde future as a given—a foregone conclusion, which we're not meant to question. . . . The sci-fi movies of the past were often utopian or cautionary; this film seems indifferent, blasé, and maybe, . . . satisfied in a slightly vengeful way."

"IS THIS TO BE AN EMPATHY TEST?"

Scott acknowledges being affected by the review: "Pauline Kael spent three pages destroying *Blade Runner* and me. . . . I couldn't believe it, it was personal. I never met her in my life and it was really distressing. . . . But after that moment, I never, ever read press again."

But it's possible to interpret, or extend, Kael's insight in a less negative way. *Blade Runner* isn't indifferent, rather it reflects *our* indifference about humankind as Earth spoiler, about this future as a given, because that, at the start of the 1980s, was how it seemed. It doesn't force passivity on us, it holds up a black mirror to our own passivity. 'We' had decided that this future was a foregone conclusion—a self-fulfilling prophecy.

Of course, *Blade Runner* wasn't the first dystopia (we'll discuss much earlier dystopias a bit later). But perhaps part of the reason for its longevity is that it's actually quite a different kind of dystopia. Normally, the purpose of dystopias is to act as a warning, so that they *don't* come true. As Frédéric Claisse and Pierre Delvenne put it: "A dystopia is the depiction of a dark future based on the systematic amplification of current trends and features. . . . Yet, however inescapable this future may be described as, the very existence of such a narrative presupposes that the political community it tries to reach is actually able to do something to thwart it. Oddly enough, a successful dystopia aims at making itself obsolete."[2]

So however dispiriting they might seem, dystopias can empower us. Claisse and Delvenne again: "In a way, dystopias operate as narrative 'time machines': bringing the future back into the present in order to change the future again." But *Blade Runner*, unusually, doesn't seem to be trying to do that. It's not a time travel movie, either (we'll discuss one of those in the next chapter). It says, this future has already happened. Deal with it.

Deckard goes to Tyrell's headquarters. He's welcomed by Rachael (Sean Young), Tyrell's beautiful personal assistant. Dr. Eldon Tyrell (Joe Turkel) appears and asks Deckard about the Voight-Kampff test, doubting its ability to distinguish Replicants from humans. Saying he wants to see a negative test result on a human first, Tyrell offers Rachael as a test subject. Deckard asks Rachael over one hundred questions. Tyrell then asks Rachael to leave. Deckard is surprised; it usually takes 20 to 30 questions to identify a Replicant. But Tyrell reveals Rachael is unaware she's not human; she's an experimental Replicant designed with implanted memories to control her—its—behavior.

Despite its dramatic failings, *Blade Runner* is effective in conjuring a thick air of alienation in which ordinary people are trapped in warrens of (over-)consumption and copies, while being redundant—economically, socially, and physically left behind. Alienation was a major theme of Philip K. Dick's work, particularly what he came to think of as "android personalities": people who had become detached from human values of compassion and empathy (this was inspired by research Dick did for his 1962 alternate history novel *The Man in the High Castle*, in which the Axis Powers won the Second World War).

Of course, to highlight this, we're meant to recognize that the android slaves, the real workers, are indeed "more human than human" (as the Tyrell Corporation's slogan has it). As we'll see, they're full of passionate intensity, still wanting to live, while the humans have given up. These are the beings that Deckard is tasked with hunting down and killing on behalf of the authorities, and because the arrogant corporation screwed up its programming and assumed there was no limit to the exploitation of its slaves.

It's a stagnant future, in the sense it keeps recurring in our visions, but also in itself: despite the flying cars and the ability to fabricate 'people', this future seems static. There's no sense of dynamism or progress here, as Kael notes, no sense of optimism that things might get better. Yes, the corporation remains intent on advancing its technologies, but it's detached from any sense of *social* advancement. The future won't be trickling down from Tyrell's gleaming corporate pyramid to the dirty, suffocating streets hundreds of floors below.

It's exactly *Blade Runner*'s anomie that helps to make it our recurring dystopia. We recognize, though don't enjoy, the resignation, because we too seem to have accepted it as our inevitable future. There will be no rebellion against the system, it's too late for that (the only dissidents are the replicants, and then only for a bit more life). The future has happened. This is why *Blade Runner* hasn't really dated. It leaps ahead of the dystopias that came before it in the 1970s and says that the revolution against the future failed, that civilizational collapse wasn't averted, and all we can do is to try to live within it (those that have no other choice).

Another thing that helps *Blade Runner*'s ubiquity as a dystopia is that its seeming absence of any explicit politics makes it unusually amenable to conservatives and liberals alike. The former can focus on the disordered urban decay, its chaotically multicultural masses and its hypocritical metropolitan elites who have long since fled, its Californian metropolis overrun, a trash heap of history. Meanwhile, liberals can focus on the inequality, public squalor, corporate power, and worker-slavery. Indeed, as *Tomorrowland* states, each side can enjoy what the other sees as dystopian. What an exciting future to look forward to!

"IF ONLY YOU COULD SEE WHAT I'VE SEEN WITH YOUR EYES"

Deckard investigates Leon's apartment, where he finds a clue: an animal scale. Leon meets up with Roy Batty (Rutger Hauer), the leader of the Replicant group, and the most advanced and intelligent of them. They go to a shop, Eye World, owned by Hannibal Chew (James Hong). He makes artificial eyes for Tyrell. Roy threatens Chew to try find out about their "incept dates"—and so when they are destined to die. Chew doesn't know, but tells them a Tyrell employee, J. F. Sebastian, may be able to get them access to Tyrell.

Artificial limbs and organs are one of the tropes of a science fiction subgenre called cyberpunk, which *Blade Runner* helped to crystallize. But it wasn't the beginning. Among other things, the film was influenced by *The Long Tomorrow*, the French graphic artist Mœbius' hard-boiled futuristic detective comic published in 1975 (written with Dan O'Bannon, who also wrote the screenplay for *Alien* and *Total Recall*, the latter based on another Philip K. Dick novel). It was a milestone in the development of the genre.

It's been a long tomorrow ever since. Four decades after *Blade Runner*, popular media continues to feature cyberpunk imagery and themes. One of the most successful videogames of the past few years has been *Cyberpunk 2077* (strapline: 'No future'), alongside other big budget titles such as *Watch Dogs: Legion* (set in a near-future post-Brexit dystopian London), and numerous independent studio games such as *Ghostrunner*, *Cloudpunk*, *The Ascent*, and *Stray*. Sci-fi television and movies also keep returning to cyberpunk, from *Altered Carbon* to *Cowboy Bebop* (originally a Japanese anime), *The Matrix Resurrections* to the Korean film *Jung-E*, not to mention numerous novels and graphic novels. A single frame from any of them instantly tells us what kind of future we've landed in: post-industrial, polluted, grossly unequal dystopias, with highly advanced technologies the playthings of the rich. It's as if the future hasn't moved on at all, which in many ways it hasn't.

This is unique in sci-fi in having avoided becoming hopelessly dated. Cyberpunk was a reaction to sci-fi's shiny spacesuits and silver rockets of the 1950s, which still persisted into the 1960s, before everything changed (we'll discuss 1950s sci-fi more in the next chapter). It was trying less to be predictive, in the tradition of so-called hard science fiction, rather it was reflective of what already was, or would be very soon. So why is it still with us?

Novelist Bruce Sterling, who helped create the genre, has said that this future is about "old people in big cities afraid of the sky." The term 'cyberpunk' was first used as the title of a short story by Bruce Bethke, written in 1980 and published in *Amazing Stories* in 1983, and was popularized by Gardner Dozois, editor of *Isaac Asimov's Science Fiction Magazine*. It's also about megacorporations and a dispiriting lack of power for ordinary people. To sci-fi author and critic David Brin, "a closer look [at cyberpunk authors] reveals that they nearly always portray future societies in which governments have become wimpy and pathetic . . . Popular science fiction tales by Gibson, Williams, Cadigan and others *do* depict Orwellian accumulations of power in the next century, but nearly always clutched in the secretive hands of a wealthy or corporate elite."[3]

Some might dismiss cyberpunk's continuing popularity as merely 1980s nostalgia, a strange kind of nostalgia about a fictional future first created decades in the past. But it still resonates because the dystopian fears of the 1980s are still our fears. As Issac L. Wheeler puts it, cyberpunk is not 1980s retrofuturism—it's still futurism, or 'presentism' really. Compared to alternative history subgenres such as steampunk, "Cyberpunk is not and never has been a retrofuturistic genre. . . . [I]t is still about the future, not a future once imagined."[4]

It is true, however, that through repetition, cyberpunk may have lost some of its power to (future) shock (we'll return to this in chapter 9). But if the future hasn't really changed . . .

"I HAD IN MIND SOMETHING A LITTLE MORE RADICAL"

So, welcome to our default future. As writer and game designer Kyle Marquis said in a well-known 2013 tweet: "Yearly reminder: unless you're over 60, you weren't promised flying cars. You were promised an oppressive cyberpunk dystopia. Here you go."

Typically, in cyberpunk, the United States has collapsed, economic power has shifted east, inequality is immutable, climate change is unchecked, and technologies of control are pervasive. Confusingly, given the genre it helped to form, *Blade Runner* lacks cyberpunk's engagement with the internet and virtual worlds. Reflecting its position as the culmination of the previous decade's sci-fi, the film is all 1970s-style chunky physical hardware.

But it shouldn't be surprising that cyberpunk feels much more like our world than most sci-fi, since it was designed that way. William Gibson's 1984 novel, *Neuromancer* (alongside *Blade Runner*, one of the foundations of cyberpunk), portrays a world dominated by a few super-automated, giant transnational electronic firms, in which most citizens live out marginal, semi-criminal existences, playing arcade games, surfing 'the Net', and taking and selling drugs. It was the future, but not that futuristic really, not in its fundamentals. Gibson's response to his fiction being regarded as dystopian was to express surprise—and skepticism: "I suspect that the people who say I'm dystopian must be living completely sheltered and fortunate lives. The world is filled with much nastier places than my inventions, . . . and a lot of those places seem to be steadily getting worse."[5]

Similarly, as Ursula K. Le Guin said, science fiction isn't predictive, it's descriptive. As suggested (by T. R. Napper), perhaps excitement about the future was already exhausted, and cyberpunk expressed why: "Where much of the sci-fi that had gone before saw technological progress as an intrinsic good, works such as *Neuromancer* and *Blade Runner* saw the dark potential of such power. . . . [Its] anti-heroes are the conflicted androids and private eyes that have seen the corruption of immense power and the suffering of the marginalized and dispossessed."[6]

"GOSH, YOU'VE . . . GOT SOME REALLY NICE TOYS HERE"

Rachael surprises Deckard at his apartment. She knows what Tyrell revealed to Deckard about her, but she doesn't believe it, and shows Deckard a picture of her with her mother. Deckard tells Rachael her childhood memories have been taken from other people. She's upset, and leaves. In another part of the city, another of the Replicants, Pris (Daryl Hannah), a 'pleasure model', befriends J. F. Sebastian (William Sanderson), a genetic designer for the Tyrell Corporation. Pris 'agrees' to stay with Sebastian in his apartment, where he lives alone and designs lifelike toys and dolls to keep himself company.

Cyberpunk critiques how technology doesn't benefit everyone equally. For the rich, advanced technologies are available for improving and extending human life; effectively, the rich can afford to live forever. Meanwhile, the poor are forced to lead a miserable existence in impoverished, ecologically polluted areas, often dying at a young age (Sebastian suffers from Methuselah Syndrome—accelerated aging—causing him to fail the medical examination to leave Earth for the off-world colonies). Technology becomes a metaphor for stark inequality.

As we'll discuss in chapter 6, cyberpunk exposes the lie of 'accelerationism', the dangerous desire to bring the super-advanced technological future closer (and faster). It conjures into being not a world of abundance and leisure for all, but hyper-wealth and power for a few, and poverty and desperation for the many.

As in 1970's *Future Shock* (a popular nonfiction book we'll discuss in the next chapter), cyberpunk depicts a world in which people are overloaded and numbed by an accelerated future. No wonder some of them retreat into chemical or virtual worlds. Of course, the technology doesn't affect everyone equally, as Craig Hickman explains: "We live in a time of both acceleration and deceleration, depending upon what end of the spectrum of the political and economic ladder one happens to exist. Those on the top tier seem to effortlessly glide through this spatial matrix with little difficulty, escaping the struggling masses of economic slaves in their zombie enclaves far below."[7]

At the same time, as Isvtan Csicsery-Ronay Jr. suggests in *The Seven Beauties of Science Fiction*, today's social, political, and cultural conditions have become 'science fictional.' Whereas once the genre was seen as inherently artistically shallow, now sci-fi imaginations help us make sense of a world being changed by proliferating technologies we—like *Blade Runner*'s inhabitants—can scarcely comprehend: nanotechnology, genetic engineering, wearable technologies, cloud computing. . . . Science fiction is the way we try to imagine the actual future, in novels (both explicitly sci-fi and not), films, TV programs, videogames, music, commercials, marketing, and beyond.[8]

One of the hallmarks of the genre is how some people defiantly appropriate these advanced technologies ("The street finds its own uses for things," as Gibson wrote in *Burning Chrome*). However, the prevailing sense remains that we can't hope to change the system. And like many other postapocalyptic stories (we'll discuss the crucial difference between collapse and postapocalyptic stories in later chapters), there's a power fantasy here as well: that in this mean future *we* would be one of the cyberpunks, finding ways to hack the matrix and turn technology against its masters for subversive ends. Not to bring down the system, however; merely to eke out spaces to survive within it.

"I'M NOT IN THE BUSINESS. I AM THE BUSINESS."

Deckard analyzes one of the photographs he took from Leon's apartment. He sees a woman hidden in the background. Along with information about the (artificial) animal scale, he tracks down another of the Replicants, Zhora (Joanna Cassidy), an assassin model who is working as an exotic dancer (exploitation again). She tries to escape, but Deckard shoots her. Bryant arrives on the scene to congratulate Deckard. He tells Deckard there are four Replicants left. Deckard insists there are only three, but Bryant tells him that Rachael has disappeared. Deckard later spots Rachael in the street. He follows her, but he's suddenly grabbed by Leon. Deckard is only saved when Rachael shoots Leon in the head.

It's not often commented on, but in *Blade Runner*, the police seem to focus on cleaning up the mess created by a corporate screw-up (meanwhile, the streets are squalidly lawless). So, we, the ordinary people, need to find our own ways to survive, because there's no state coming to protect us. Gibson has explained that in his *Neuromancer*, following a catastrophic war: "I wanted to write a novel where multinational capital took over, straightened that shit out, but the world was still problematic."[9] In our world, we didn't need a nuclear exchange to lead to the retreat of the state from ensuring the health and security of citizens. In our world, as we'll see, we had an ideology that successfully argued for this on its own, and that multinational capital and corporate power would 'straighten shit out' for us. In cyberpunk, corporations become the law; in our world, that was achieved by some highly partisan economists and corporate lawyers, some of whom, it seems, would like us to go full cyberpunk, with untrammeled corporate power, private police forces and legal systems, and no meaningful welfare safety net at all.

Rachael asks Deckard whether he'd hunt her down if she disappeared 'to the north.' He tells her he wouldn't—but someone else surely would. She asks about her incept date, but Deckard says he didn't look at it. Meanwhile, Roy comes to Sebastian's apartment. He reveals to Sebastian that he and Pris are Replicants, and they have a similar condition to Sebastian: accelerated aging. Roy coerces Sebastian into arranging a personal meeting with Tyrell. Their time is running out.

In theory, this should be an age of flourishing optimistic visions of future societies, not necessarily because everything is perfect, but because it's *not*, because the challenges and crises we face could be motivating artists, designers, and planners to imagine how we might respond.

Instead, we're pessimistic about the future, and for good reason. Americans, among others, see a country with colossal national debt, a widening gap between rich and poor, elite detachment and corruption, and workers threatened by seemingly 'unstoppable' forces of outsourcing, contract labor, and automation. And that's before we get to climate change and ecological collapse. Majorities predict that the economy will be weaker, healthcare will be less affordable, families will find it harder to make ends meet, and the environment will get worse. Most think that the economy benefits the rich and hurts the poor.

All of this is rather perplexing to techno-optimists like Steven Pinker, the Canadian-American cognitive psychologist, linguist, and popular science writer, who urge us to consider how historically things are continuously getting better. As Erik Olin Wright (not a techno-optimist) sums up this bemusement: "For many people, the idea of anticapitalism seems ridiculous. After all, look at the fantastic technological innovations in the goods and services produced by capitalist firms in recent years."[10] But as cyberpunk captures, many people feel that technological progress, and the corporations that sell it to us, have turned against them. (This goes back to Philip K. Dick's work, whose stories are concerned with the dehumanizing effects of 'progress.') Whatever its wonders, technological advance doesn't seem able to solve the social and environmental problems we face, indeed there is the increasing sense that it might be *causing* them.

"NO CHOICE, PAL"

More broadly, capitalism may have lost its power to create exciting but credible positive visions of the future for more and more people, and is leaving us with a growing list of crises. Nonetheless, the free-market utopia we were promised still retains a residual power in the ideas of opportunity for those who work hard, along with competition, innovation, scientific breakthroughs, and endless wealth creation—a world increasingly knitted together by commerce, culture, and technology. But the obvious answer to the optimists is that this is also an age of poverty, inequality, insecurity, elite power, climate change, and environmental destruction—a future that cyberpunk says is on its way, or is here already (hence the genre's well-known description as a "combination of low-life and high tech"[11]).

The more fundamental question, and the central theme of this book, is why we don't feel we can change these things, which is really a *political* question (a political question about politics). As we'll discuss, this is the real underlying *early collapse* of our societies, and it's the outcome of a particular political project. It's also why a cyberpunk future seems so recurringly relevant.

For the theorist Mark Fisher, the prevalence of dystopia visions in capitalist societies demonstrated how incapable we are of constructing viable alternatives to capitalism. But that's not quite true, as we'll see. It's really more practical and more obvious than that. The decades-long persistence of this particular vision of dystopia reflects our collective recognition that this future is unavoidable. It's not 'what if?', it's 'are we there yet?'

For example, recently there's been a trend, on social media in particular, of noting the resemblances between dystopian fiction and reality, for example September 2020's eerie images of San Francisco's skyscrapers against a dirty orange sky caused by pollution and wildfires drew numerous comparisons with scenes from the sequel *Blade Runner 2049*. Nevertheless, our lagging economic and political systems remain largely in denial, and like any denial, this causes depression, a stuckness, not only in individual depression but also in a collective social depression—a condition of the dystopia we're in already.

Roy and Sebastian take the elevator to Tyrell's penthouse atop the corporate headquarters. In Tyrell's bedchamber, Sebastian introduces Roy. Tyrell seems to have been expecting him. Roy confronts him; he wants his lifespan extended. But Tyrell says it isn't possible. Despondent, Roy first kisses Tyrell, then crushes his skull in his hands. After Tyrell's murder, Deckard searches Sebastian's apartment. In a spooky scene, Pris, hiding as a mannequin, suddenly attacks him, but Deckard manages to shoot and kill her.

In another respect, *Blade Runner* was really 'just' an updated version of an urban nightmare established decades earlier and revived in the economic and social context of the 1970s. The granddaddy of futuristic urban visions is Fritz Lang's towering *Metropolis*. Set in 2026, it's a comprehensively realized depiction of a city with no apparent middle class. The ruling elite class, called the Club of the Sons, are led by a ruthless industrialist. They've separated and subjugated the rest of the population as slaves, who live deep underground below the factories where they work, men and women who've been condemned to keep the machines of the city above them running smoothly.

For much of the twentieth century, the vertical city became linked with dystopian sci-fi visions. Following *Metropolis*, this literal depiction of class hierarchy has been richly portrayed in a variety of sci-fi stories, from J. G. Ballard's *High Rise* (1975), to more recent stories such as *Elysium* (2013) and *Altered Carbon* (2018). In *Blade Runner* in particular, the vision of an immense, soaring, efficient planned city was subverted into a sprawling, broken, dilapidated, under-invested near-slum—this in the city that has so often represented the future, Los Angeles (we return to public decay in the next chapter).

One reading of cyberpunk cities is then that they depict a profound disinvestment in the public sphere, in infrastructure, and in the future. The shiny soaring cityscapes of the 1950s future were invested in and coordinated by *somebody*, after all, as well as reflecting the growth and general optimism of the era. They also seemed to be shared futures; the promise was that everyone would benefit from them (at least in theory). Not anymore.

Blade Runner's neo-noir atmosphere and tropes are well-chosen. As a genre, noir is linked to economic and social depression, cynicism about the present (elites get away with their crimes and corruption, while the little guy never wins), and pessimism about the future.

Cyberpunk's addition of advanced technologies serves to reinforce this pessimism: we're surrounded by wonders, but there's no shared sense of social excitement. In noir, there's an unease, to say the least, with modernity. And *Blade Runner* is noir resurfacing in a 1970s context (extrapolated into the future) – once again, an era marked by decay, alienation, and a lurking sense of catastrophe. From a noir perspective, the film has another precursor in *Alphaville*, the 1965 French New Wave sci-fi neo-noir directed by Jean-Luc Godard. That film follows a secret agent in a city of tomorrow (although it was shot in real modernist locations in Paris), a technocratic, emotionally repressive dictatorship managed by a dictatorial computer, Alpha

60. Eventually, the agent and the regime leader's daughter escape the city. As we'll see, however, our ending is likely to be far less happy.

"WHAT DO YOU MEAN, I'M NOT HELPING?"

We can go back much further. Dystopian fiction stretches back to the eighteenth century. Mary Shelley's 1826 novel *The Last Man*, set mainly in the 2090s, established many of the tropes of dystopian fiction that are still employed today: continual conflict, rampant elitism, downtrodden societies, and a dying humanity (the story describes a future Earth that's ravaged by a pandemic, ultimately resulting in the near decimation of humanity).

But Shelley's warnings were ignored, as were those of later authors such as Jack London, who penned *The Iron Heel*, published in 1908, another early dystopian novel. Instead, fiction generally portrayed better tomorrows due to the ongoing scientific revolution, bringing technological breakthroughs that cure many human woes and make life better for many.

This optimism finally began to wane in the late 1940s. The uncertainty left by the Second World War, along with the horrors of potential atomic annihilation, prompted fiction to adopt darker themes once again. This period saw the beginnings of postapocalyptic stories that would become a staple of the Cold War era. But as we'll see in chapter 4, dystopianism really resurfaced in the 1970s. And *Blade Runner*, as noted, can really be regarded as a 1970s film, in many ways the culmination of that decade's dark turn in sci-fi, as can Ridley Scott's previous film *Alien* (1979), with its similar themes of corporate exploitation and unchecked desire for technological dominance, alongside its dirty future production design (fans even like to speculate that the two film franchises are set in the same fictional universe).

Given this turn away from bright optimism and belief in science and progress, we have to ask whether the prevalence of fictional dystopias is helpful, especially when we face so many major global challenges. Even worse, cyberpunk's hacker vibe and funkily exotic, but still believable, near-future environments might even make social, political, and environmental collapse seem superficially, momentarily, kind of *cool*. Maybe worse than resignation, could it be that we're looking forward to this end of the world?

As noted, the conventional purpose of dystopias is to act as a warning. Margaret Atwood, perhaps the most celebrated contemporary author of dystopian fiction, whose 1985 novel *The Handmaid's Tale* takes place in a near-future strictly patriarchal United States, also suggests they can prompt us to think longer-term: "If you're interested in writing speculative fiction, one way to generate a plot is to take an idea from current society and move it a little further down the road. Even if humans are short-term thinkers, fiction can anticipate and extrapolate into multiple versions of the future."

But what value is there to repetition when, as *Blade Runner* assumes, we didn't listen when there was still time? Echoing Kael, some commentators suggest that the recurringly bleak futures of contemporary sci-fi are potentially regressive, that once shock futures have lost their power, we are left only with apathy.

In this vein, historian Jill Lepore identifies a "radical pessimism" in contemporary dystopian fiction: "Dystopia used to be a fiction of resistance; it's become a fiction of submission, the fiction of an untrusting, lonely, and sullen twenty-first century . . . It cannot imagine a better future, and it doesn't ask anyone to bother to make one."[12] Indeed, cyberpunk might have inadvertently served to normalise elements of dystopia—gross inequality, corporate control, environmental collapse—we should be working against. And so, endlessly navigating the crowded streets of cyberpunk's maze-like cities only takes us further away from imagining the newer, better worlds that could be.

Fredric Jameson begins his essay in "An American Utopia: Dual Power and the Universal Army" (2016) with the observation that, "We have seen a marked diminution in the production of new utopias over the last decades (along with an overwhelming increase in all manner of conceivable dystopias, most of which look monotonously alike)." Jameson recognizes how capitalism continually reproduces dystopias. Risk-averse producers invest millions in tried-and-tested, homogenizing dystopias. In the same book, Slavoj Žižek insists that the "dystopias that abound in recent blockbuster movies and novels (*Elysium*, *The Hunger Games*), although apparently leftist (presenting a postapocalyptic society of extreme class divisions), are unimaginative, monotonous, and also politically wrong." If images of the future are important, including politically, shouldn't we focus on creating and consuming more positive, or at least more balanced, futures (since as we'll see, there are some)? Echoing these critiques, some green commentators and campaigners bemoan real-world environmental doomism as disempowering and self-fulfilling.[13]

"CAN THE MAKER REPAIR WHAT HE MAKES?"

Such sentiments are hardly new. Dutch historian and sociologist Fred Polak argued in his 1973 book *The Image of the Future* that we're now in a "timeless time" because of the "de-utopianising of utopia." (Polak was a former managing director of the Netherlands' Central Planning Bureau and one of the founding fathers of futures studies.) To Polak, we are "future-creating beings," citizens of two worlds, the present and the imagined.

Polak traces visions of the future through history, through religious traditions and 'secular religions' such as Marxism. To Polak, now however our images of the future are nihilistic and full of despair, they are visionless and reflect a "denial of any power" to create anything new. Instead, utopianism is the forerunner of all social policy and organization: "The decline in the utopian images of the future reflects a lost faith in human self-determination in our own time. . . . [W]e may well ask ourselves if the decline in utopian thinking is not also a decline in social progress itself."[14] To Polak, a generation of young men and women desperately need to know that it is possible to create other and better futures.

Or maybe counterintuitively, there's also the notion, as Sam Kriss says, that: "Dystopia lets the present off the hook: it takes all the worst things about actually existing society and displaces them into the future; horrors that really do press on us from all sides instead become an imaginative fantasy."[15]

Perhaps worst of all, a popular culture dominated by dystopian visions in which survival is depicted as individualist, violent, and ultimately hopeless, not only helps to bring these worlds into being, but it also serves the interests of the elites who will be the only ones to live securely and comfortably in these worlds, ensconced in their towers or behind heavily guarded gates.

But then again, as T. R. Napper suggests about noir (which equally applies to cyberpunk), perhaps this misunderstands what these genres are trying to do: "Noir is not designed to map out a utopian future, but to reveal the nature of an unjust present. Noir is concerned with modernity's casualties, those left behind in the creative destruction of capitalism [and] technological revolution."[16]

In this vein, Tom Moylan (in *Demand the Impossible*, 1986) proposed a now well-known distinction between anti-utopias and dystopias. Anti-utopias are just that: arguments against the possibility of utopia, a warning and denunciation against utopian programs based on the conviction that human nature is so inherently corrupt it could never be redeemed by a better social order (we'll discuss anti-utopianism later in this book). In contrast, dystopias are a contradictory, hybrid form, despairing but not without hope.

What's commonly misunderstood is that dystopias aren't the opposite of utopias, they're a kind of negative cousin. To explain: in *Scraps of the Untainted Sky* (2000), Moylan developed the idea of 'critical dystopias.' These, he explained, "burrow within the dystopian tradition," but do so only "in order to bring utopian and dystopian tendencies to bear on their exposé of the present moment." They are "stubbornly" utopian, in the sense that they do not move easily toward their own better worlds: "Rather, they linger in the terrors of the present even as they exemplify what is needed to transform it."

So, what Pauline Kael was really reacting to was the lack of *critical* dystopianism in *Blade Runner*—the sense it's too late to transform this world, and it's been too late for some time.

"ALL I COULD DO WAS SIT THERE AND WATCH HIM DIE"

Roy arrives at Sebastian's apartment. He finds Pris dead, and weeps. He hunts Deckard around the empty building. Deckard tries to escape by jumping to another roof, but barely makes it and clings on by one hand. Roy makes the jump easily and peers over the edge at Deckard. Deckard slips—but Roy catches him, saving his life.

Roy sits down, dejected. He tells Deckard about the wonders of the universe he's seen, lamenting that his memories will now be lost, "like tears in rain." Then he dies.

Criticizing this type of dystopia amounts to blaming creators for *society's* stasis. Before we accuse them of being complicit in creating pervasive social pessimism, we should (much more importantly) consider our own complicity in the systems that are leading steadily towards dystopias, and how it's these systems that are the cause of hopelessness—not novels, movies, comics, or videogames. It's not a failing of creators' imaginations that we're stuck in dystopian visions of the future; it's that politically, economically, and socially we're stuck on the same trajectory toward a future that's dystopian (in many ways, is already dystopian, as we'll see). As Mark Bould puts it: "We already inhabit the worst of all possible worlds—the one that actually exists—so perhaps there is no critique left that dystopia can effect. Perhaps, reduced to a spectacular commodity, to obscene surface, it has nothing left to tell us that we do not already know."[17]

But maybe that's the point. 'The Future' never changed because the new right that rose to power in the 1980s worked to make sure that it didn't. So it's hardly surprising that we're still being presented, and still consuming, *Blade Runner*–like futures, because according to the new right, the deregulated, corporate-dominated, state-absent future was the only future possible anyway. No wonder the only future that we seem able to imagine is an extension of the present. No other futures are permissible.

If there's any perverse silver lining, it might be that since fictional dystopian futures are increasingly materializing in our real world, the more we recognize this, the more we might finally be able to move on to considering what kinds of utopias we might be able to create out of the crumbling city collapsing around us. That's what this book is about.

Finally, there's another consideration we'll get into later in this book, that some dystopias are also fallen utopias—schemes for perfect societies gone horribly wrong. But what could the hoped-for, once-future utopia possibly have been in the case of *Blade Runner*?

The answer is so obvious that it's perhaps difficult to see, because we're in it. It's industrial capitalist civilization, and even more, the robust defense and promotion of its virtues by the new right, especially from the 1970s. Despite their proclaimed distrust of utopian schemes (which they saw as inherent to liberal-left politics), this new right had its own utopianism: a perfect order, organized in accordance with the 'free market', which if allowed to accelerate would provide for every need and create the future seemingly on its own.

The grimy littered, acid rain splattered, poverty-stricken streets of *Blade Runner*'s Los Angeles 2019 put the lie to that paradise, as do many of our own streets. But even while it's a dystopia for the many, it's a utopia for some. It's a good place for the corporate titans protected atop their pyramidal towers (at least until their creations come back for them), and for those who were wealthy and healthy enough to be able to leave the planet for the off-world colonies—just like our own transnational, insulated elite classes.

This is the hidden-in-plain-sight critique that cyberpunk continues to convey, even as we can't avoid its future crashing into our present: there was no sudden shock, no asteroid or solar flare, that caused this social collapse, just the refusal to change course even after we were warned that economic growth, extraction, and exploitation couldn't go on forever.

"IT'S A SHAME SHE WON'T LIVE. BUT THEN AGAIN, WHO DOES?"

Gaff appears and congratulates Deckard on completing his investigation. Deckard says he's done with hunting Replicants. As Gaff walks away, he makes his flippant comment about Rachael's lifespan. Deckard returns home, and finds Rachael waiting for him.

Notoriously, the ending of the original theatrical version of *Blade Runner* was forced on Scott by the studio (the film had tested awfully with audiences). Scott complied, wanting his movie to be a success (he also lacked final cut privileges). It had ended with Deckard and Rachael fleeing his apartment, cut to black. Now it concluded with tacked-on, tonally discordant shots of them driving through unspoilt countryside (strangely enough, the very first version of *Blade Runner*'s script ended with an extended pullout shot that would bring the blue-green sphere of the Earth into view). The new ending was accompanied with more virtually monotone voiceover from Harrison Ford (who hated the changes): "Gaff had been there, and let her live. Four years, he figured. He was wrong. Tyrell had told me Rachael was special. No termination date. I didn't know how long we had together . . . Who does?"

The couple fleeing the dark city for freedom and fresh air is a staple of dystopian fiction, but it doesn't work here. As is well known, the landscape shots were unused footage from Stanley Kubrick's *The Shining*. Even if you don't know this, they seem like they're from another film. In any case, cyberpunk is not about happy endings.

Blade Runner is notable for its different released cuts, *seven* in total, although only three are really acknowledged by fans: the theatrical cut; the director's cut; and the final cut. The later versions, which lose the happy ending, also point towards something that was only lightly hinted at in the original: the implication that Deckard is probably a replicant.

The Tyrell Corporation implants false memories into its creations in order to 'stabilize' them. In a new scene added to later versions, Deckard dreams of a unicorn galloping through a forest. We never see him telling anyone about his dream. But in the final scene, it's Gaff, who we've previously seen creating little origami animals, who has presumably left the silver paper unicorn outside Deckard's apartment. Scott even confirmed the implication in the short documentary *Deck-A-Rep: The True Nature of Rick Deckard*, stating, "How would anyone have known what was inside his head other than someone who knew what was in his file, that had been implanted in his brain? Can't be any clearer than that."

But *Blade Runner*'s themes run deeper than arguing about the nature of its protagonist. Rather than debating whether Deckard is an android, we're surely supposed to question the difference between the Blade Runner and the Replicants he's tasked with disposing of. Relatedly, the film also forces us to question what it means to exist in an alienatingly bleak near-future of industrial smog, towering Coca-Cola advertisements, and urban loneliness.

"MAYBE IN THOSE LAST MOMENTS HE LOVED LIFE MORE THAN HE EVER HAD BEFORE"

Blade Runner reminds us that we'll all die someday, true, there's no escape from that. But also that to live in fear is to be a slave, Replicant or not. Politically, environmentally, socially, the film points in the same direction. We've failed to escape the future, it was all decided a long time ago. There is no escape from the city into a pristine wilderness. To imagine a happy ending is a fantasy; it makes no sense from what's come before.

To Pauline Kael's criticism then, we can say that the film's fatalism is not its flaw, it's its point. Ultimately, dystopia comes from a political collapse, an inability of politics to respond to oncoming crises, and far from being accidental or some collective human failure, this inability was core to the political project of the new right, as we'll see.

In response, politics on the edge of dystopia needs to recognize how things have changed forever. The future is set, or at least collapse is, both by what people believe is possible (informed by forty-plus years of new right politics), and the long-term trends put in place largely by their policies, which are still playing out (carbon emissions, resource extraction, de-unionization, inequality, people's vulnerability, automation, outsourcing . . .). They've both defined the future, and how we feel unable to envisage any other possible future.

Once we recognize that collapse is on its way, once we accept that fictional depictions are coming true, it changes not only our politics but also our personal lives. It changes everything. We won't be able to flee the city, but we do need to find ways to survive within it. And by doing so, perhaps rescue our own humanity as well.

NOTES

1. Pauline Kael, "Baby, the Rain Must Fall," *The New Yorker*, July 12, 1982.
2. Frédéric Claisse and Pierre Delvenne, "Building on Anticipation: Dystopia as Empowerment," *Current Sociology* 63, no. 2 (2014).
3. David Brin, *The Transparent Society* (New York: Basic Books, 1998).
4. Issac L. Wheeler, "Cyberpunk Is not Retrofuturism," *Neon Dystopia*, August 28, 2016. www.neondystopia.com/cyberpunk-politics-philosophy/cyberpunk-is-not-retrofuturism/.
5. David Wallace-Wells, "William Gibson, The Art of Fiction No. 211," *The Paris Review* 197, summer, 2011.
6. T. R. Napper, "A Post-Pandemic World: The Ultimate Cyberpunk Primer Discussions," *Grimdark Magazine*, May 20, 2020. www.grimdarkmagazine.com/a-post-pandemic-world-the-ultimate-cyberpunk-primer/.
7. Craig Hickman, "Fredric Jameson—Utopia, Dystopia, and the Myth of Neoliberalism," Copyriot.com, September 28, 2016. https://non.copyriot.com/fredric-jameson-utopia-dystopia-and-the-myth-of-neoliberalism/.
8. Istvan Csicsery-Ronay Jr., *The Seven Beauties of Science Fiction* (Middletown, CT: Wesleyan University Press, 2008).
9. David Wallace-Wells, "William Gibson, The Art of Fiction No. 211," *The Paris Review* 197, summer 2011.
10. Erik Olin Wright, "Why Be Anticapitalist?," Verso Books blog, August 26, 2020. www.versobooks.com/blogs/4419-why-be-anticapitalist.
11. As coined in Bruce Sterling, "Preface," in *Burning Chrome*, William Gibson (London: Harper Collins, 1986), xiv.

12. Jill Lepore, "A Golden Age for Dystopian Fiction," *The New Yorker*, June 5 and 12, 2017. www.newyorker.com/magazine/2017/06/05/a-golden-age-for-dystopian-fiction.

13. For example, Sasha Lilley, David McNally, Eddie Yuen, and James Davis, *Catastrophism: The Apocalyptic Politics of Collapse and Rebirth* (Oakland, CA: PM Press/Spectre, 2012).

14. Fred Polak, *The Image of the Future* (London: Elsevier, 1973), 179.

15. Sam Kriss, "Abandon the Future," *Jacobin*, November 21, 2015. www.jacobinmag.com/2015/11/back-to-the-future-day-marty-mcfly-sam-kriss.

16. T. R. Napper, "Kim Stanley Robinson Is Wrong About Something," *T.R. Napper*, April 5, 2019. www.nappertime.com/kim-stanley-robinson-wrong-something/.

17. Mark Bould, "Dulltopia: On the Dystopian Impulses of Slow Cinema," *Boston Review*, January 22, 2018. http://bostonreview.net/literature-culture-arts-society/mark-bould-dulltopia.

Chapter 2

"I Mean, How Can All This Be Happening? It's Like We're in Hell or Something"

How Back to the Future *Predicts the Decline of Middle-Class Security*

When we ask 'what happened to the future?,' we might think first of all about the near-magical technologies that we (older generations, at least) were seemingly promised: flying cars, jetpacks, even hoverboards. This focus on technology is understandable. But there's something more important underlying these visions which also hasn't happened: a shared increasing prosperity, ever-increasing leisure time facilitated by new technologies, and investment in shiny, efficient cities. Most of all, we're missing the optimism founded in ongoing social progress. It's the decline of this optimism that the *Back to the Future* trilogy (beginning in 1985) brilliantly depicts, and even more, that the middle-class American promise of utopia may always have been a lie, and its decline has sown the seeds of its coming (and increasingly present) dystopia.

"I NEED TO BORROW YOUR . . . HOVERBOARD?"

Marty McFly (Michael J. Fox) might seem like the average small-town, middle-class American teenager, but his life is deeply underwhelming. His family is . . . well, we'll get to them in a moment. He's disengaged by school, hassled as a loser by the school principal, and only seems motivated by his (very remote) dreams of becoming a rock star. He does however have a crazy scientist friend, Dr. 'Doc' Emmett Brown (Christopher Lloyd), who tells Marty to meet him at Twin Pines Mall at 1:15 a.m.

To Marty's amazement, Doc unveils a time machine that runs on plutonium built into a DeLorean. Unfortunately, Libyan terrorists (it was the 1980s) want their plutonium back, and as he tries to escape, Doc is shot and Marty accidentally activates the time machine, sending him back to 1955, where he inadvertently prevents his parents from falling in love. Now he'll have to find a way to get his parents together again, convince the 1955 Doc he's from the future, find a way back to his present, and save the 1985 Doc from being killed.

It might seem strange to go from the darkness of *Blade Runner* to the apparent lightness of the *Back to the Future* trilogy. But *Back to the Future* is especially smart, as well as entertaining, about how the past can lead to a dilapidated, dystopian future, and much more besides.

Arguably, only *Back to the Future: Part II* is really science fiction, although the trilogy as a whole hinges on the genre trope of an advanced new technology, in this case time travel, that becomes a hazard with all sorts of unintended consequences we have to scramble to fix. But that's just a plot device. Actually, all of the films are about utopia and dystopia, not always in the obvious way of films like *Blade Runner*, but in important ways nonetheless.

When we ask, 'what happened to the future?', we might at first think about the near-magical technologies that we (older generations, at least) were seemingly promised: flying cars, jetpacks, and yes, hoverboards. This focus on technology is understandable, as obviously visible signs of progress. But as we'll discuss, there's something more important underlying these technological visions which also hasn't happened: a shared increasing prosperity, ever-increasing leisure time facilitated by new technologies, investment in shiny, efficient cities, and a dynamic but *organized* future.

Most of all, we're missing the optimism founded in ongoing social progress. It's the decline of this optimism, among other things, that *Back to the Future* brilliantly depicts.

"THINGS HAVE CERTAINLY CHANGED AROUND HERE"

Back to the Future (the first film), directed by Robert Zemeckis and written by Zemeckis and Bob Gale, was released at the height of Ronald Reagan's America (Reagan had won reelection in a landslide the year before). Its star was a clean-cut icon of middle America, best known for playing an ardent young Republican on the popular TV sitcom *Family Ties*. The first *Back to the Future* went on to become the highest-grossing movie of the year, and the trilogy seemingly revolves around the material success of a regular suburban nuclear family.

The conclusion of the first film, and the whole series, is that the family's many dysfunctions are corrected when Marty eventually gets back to the mid-1980s. The family's comfortable middle-class status is resurrected. What could be more Reaganite than that?

But really, the films not only critique this middle-class entitlement, they also suggest that greed, selfishness, and materialism are the problem—they're what's screwed up the future—and absent the fantasy of time travel, it's too late to change what happens next. Even more, the middle-class American promise of utopia may always have been a lie, and its decline has sown the seeds of its coming (and increasingly present) dystopia.

In the first film, Marty's middle-class suburban life is actually a hollow, desperate thing. In between the sharply observed jokes, it's a surprisingly dark depiction of people who are just about hanging on. His parents live in their own worlds. His mother (Lea Thompson) is an alcoholic, his father (Crispin Glover) is being bullied by the thuggish Buford 'Biff' Tannen (Thomas F. Wilson), his brother and sister are unsympathetic underachievers, and Uncle Joey is in jail (missing parole yet again).

Their Californian town of Hill Valley ('A Nice Place To Live') also becomes less pleasant the closer we look: littered and fraying, the decline visible around the edges creeping ever inward. For one thing, the clock tower hasn't been repaired for ages, and the courthouse is in disrepair. In between the loan shops, burger joints, bars, tattoo parlors, secondhand stores, and empty shopfronts, there's an adult bookstore and the local movie theater is showing porn films. As in many towns, the mall has drained its downtown of life.

When Marty travels back to 1955 in the time machine, accidentally preventing his parents getting together threatens his own (and his brother's and sister's) annihilation. Of course, he manages to save his future. Even better, it seems, when he returns to 1985, he discovers he's changed his family's future as well. Marty's parents are successful and attentive, his father is a published sci-fi author, his mother is thin and glamorous, his brother and sister are striving professionals, and Biff is reduced to a sycophantic helper.

"A COLORED MAYOR?! THAT'LL BE THE DAY"

It's not clear that Pauline Kael really got it when she wrote about *Back to the Future: Part I* that, "the film's idea of happiness in the eighties—with one of Marty's siblings turned into a Yuppie and the other into a deb, and his parents strolling like lovers amid the pastel sofas in a living room that looks like a commercial for a furniture-warehouse sale—should be a satirical joke but isn't." Oh, but it surely *is* satire—of an idealized 1950s (which Kael also didn't get), and also of middle-class material aspirations (overall though, Kael thought the film was ingenious and likeable, to a degree).

To the first point, as we get to see, the 1950s is not the utopia sold to Americans by Reagan and the new right (later, we'll discuss what was always fantasy about this utopia). Beneath the surface of its almost spotless, vibrant downtown, this 1950s (and the real-life 1950s) is repressive, racist, sexist, violent, and hypocritically moralistic (which, when you think about it, might actually make it a utopia to some conservatives).

To the second point, because of, rather than in spite of, their failings, the earlier version of Marty's family is more likeable and relatable, more human, compared to the go-getting, positive-thinking yuppy versions he finds back in his changed present. The story slyly asks: is this 1980s superficial success all there is, is this all we really want from the future?

Indeed, is it ultimately really American utopia resurrected? As we see in *Back to the Future*, from the beginning there's middle-class status materialism and insecurity, domination and bullying, sexism, and racism—all of which develops into a future that's increasingly broken and dystopian. But it was there from the start.

The zingy coda to the first film (which was never intended to set up a sequel) is that Marty and his girlfriend Jennifer need to travel to the future with Doc to fix what's "gone wrong" with their family-to-be ("It's your kids, Marty! Something's gotta be done about your kids!"). The anxieties of maintaining middle-class security and respectability never end.

But even worse, as we'll see in a moment, this causes Marty to bring into being a dystopia by trying to perfect his nuclear family in the future. And so not only are conservative social politics—keeping the nuclear household together as a guarantee of financial security and happiness—clearly not the answer to the flailing McFly family in the 1980s, but they're also equally unlikely to be in the future either. Indeed, they cause further disaster.

The trilogy's clever title—our present is of course 'the future' that Marty is trying to get back to and correct—reflects this more critical interpretation. (Famously, studio executive Sid Sheinberg suggested changing the title to 'Space Man from Pluto.' Gale and Zemeckis didn't know how to reject Sheinberg's 'suggestions' without risking his ire, but executive producer Steven Spielberg intervened, sending Sheinberg a memo that read "Hi Sid, thanks for your most humorous memo, we all got a big laugh out of it, keep 'em coming." Sheinberg never mentioned the idea again.) Even more smartly, the title alludes to how our present was once the future, but not anymore, because it isn't what we were promised, what once seemed possible, even inevitable.

"THE FUTURE. UNBELIEVABLE. I'VE GOT TO CHECK THIS OUT, DOC."

So it's off to 2015. *Back to the Future: Part II* and *Part III* were also directed by Zemeckis and co-written by Zemeckis and Gale. Like *Blade Runner*, the whole of *Back to the Future* now takes place in our past. At first, it appears like the anti–*Blade Runner*, a technological utopia. And yet, as with its takes on 1980s suburban materialism and 1950s 'innocence', *Back to the Future: Part II*'s depiction of 2015 is (again) really a satire of it, of technology-focused speculation, including 1950s-style optimistic invention-focused futurism.

Watched again, it indicates something crucial. The 2015 future being satirized is not a 1980s vision of the future but a *1950s* one, because, as we discussed in the previous chapter, there is no 1980s (or later) technologically and socially progressive vision of the future, just *Blade Runner* playing endlessly on a loop, and we've already seen that movie many times over.

Of course, this didn't stop the articles about what *Back to the Future* got right and wrong ('where are my self-lacing Nikes?' etc.—as if that's what really matters). The filmmakers did do some research into what scientists and technologists thought might happen by 2015, but with a full awareness of the likelihood of making inaccurate predictions. Recognizing this, as Gale says, "We knew we weren't going to have flying cars by the year 2015, but God we had to have those in our movie." (Similarly, there's that retro staple of the future—weather control.) The producers even continued their satire into the real world, claiming that the hoverboards featured in the film were real but not yet available to the public due to "safety concerns" (inevitably, people still tried to order them at toy stores).

Nonetheless, they do pretty well with their 'predictions.' There are surveillance cameras, flying drones, flat-panel TVs, video calls, smart home systems, augmented reality goggles, wearable technology, tablet computers with fingerprint scanners, and payment via portable devices. But unlike in most future-set stories, and underlining its satirical intentions, the technology often doesn't work (and in a nice touch, fax machines are still everywhere).

The Hill Valley of 2015 has at least received some urban investment. Downtown has been spruced up a bit (perhaps Mayor Goldie helped) and there's an (unnamed) female president, but there are suggestions of lurking dystopia as well. Seemingly, the rule of law has been abolished (no lawyers!) in place of summary justice. And there's a throwaway reference to how much less money is worth (50 dollars for a Pepsi!), which suggests that some kind of economic calamity has happened that's caused rampant hyper-inflation.

But according to Gale, "What's wrong is due to who lives in the future as opposed to the technology." The McFlys are back to being tortured by life.

Griff (Biff's grandson) is bullying the hapless Marty Jr. The 2015 version of Marty has given up on his dream of becoming a rock star after a car accident; later he's fired from his job. And 1985 Marty and Doc have to resolve a problem with Marty Jr., who's about to trigger a family-destroying chain of events by committing a robbery. Apparently, material security, let alone cool gadgets, haven't stopped dysfunctionality from resurfacing in the McFly family. But their hopelessness really symbolizes something else—something social.

"RADIATION SUIT? OF COURSE. BECAUSE OF ALL THE FALLOUT FROM THE ATOMIC WARS!"

To understand how the future ended, we need to go back to when the future was still The Future. If cyberpunk marked the end of futurity, then this is where its contemporary version really began.

As reflected in popular culture, the 1950s and early 1960s was an age of popular futurism, backed by public investment in society and technology, most obviously in the space race between the United States and the Soviet Union, but much more broadly as well. The Future was a popular thing, in mainstream magazines, radio, TV, and advertising. It was about fantastic new modes of transport, an age of (imagined) 'flying saucers for everybody!', but it was also domestic, in time-saving homes and lifestyles.

As Kurt Anderson puts it in his *Evil Geniuses: The Unmaking of America: A Recent History* (2020), this was the age of 'peak new' – suburbs, television, rock 'n' roll music, shopping malls, and plastic everything! America did indeed feel like Tomorrowland. *Back to the Future: Part I* captures this optimistic age nicely, with its futuristically finned cars and 'space age' TV dinners. And at the strictly distanced high school dance (in *Part I*, revisited in *Part II*), Marty even takes rock 'n' roll from the future and introduces it to the past (in the 1950s, rock music must have seemed like it came from the future; here it actually did). Whereas, in the "actual" fictional future of 1985 and 2015, the promised life of domestic leisure served by amazing technological progress has dissolved into harried domestic chaos and gadgets that don't work and no-one has any sense of wonder about anything anymore.

Naturally, this futurism was reflected in, and promoted through, science fiction. But the flipsides of such progress were also questioned, often through metaphors of outward threats and progress that goes too far too fast (we'll come back to this idea in a moment). Hence the popularity of alien invasion and monster films in the 1950s, many low budget, but among them some classics, including *The Day the Earth Stood Still* (1951), *The Beast From 20,000 Fathoms* (1953), *It Came From Outer Space* (1953), *The War of the Worlds* (1953), *Creature From the Black Lagoon* (1954), *Forbidden Planet*

(1956), *Invasion of the Body Snatchers* (1956), and *The Incredible Shrinking Man* (1957).

Aliens appeared, or creatures were created, by man's (it was always man's) ignorance in messing round with forces beyond his capacity to grasp. Given the development of the atomic bomb, visitors from other planets often took the form of superior beings who challenged us to deal maturely with the power we'd attained, nicely captured in *Part I* in the scene in which a disguised Marty exploits his young father's obsession with sci-fi: "Silence, Earthling! My name is Darth Vader. I am an extraterrestrial from the planet Vulcan!" At issue in 1950s sci-fi was how we handle amazing technological progress and develop the greater sense of responsibility the future requires.

"I FINALLY INVENT SOMETHING THAT WORKS!"

Back to the Future cleverly homages this golden age of science fiction. The whole thing is built on a 1950s-style sci-fi premise, consciously so: a brilliant but crazy inventor creates a device that threatens existence (the McFly family's existence, at least). And then of course we travel back thirty years and he actually *is* a 1950s mad scientist.

As we discussed in the previous chapter, in our age dystopian stories have been accused of promoting passivity. But this criticism isn't new; 1950s sci-fi films were critiqued for doing the same. One of the first serious analyses of them was Susan Sontag's famous 1965 essay on Cold War–related science fiction fantasies, *The Imagination of Disaster*. Sontag noted astutely that 'fantasy' can both provide an escape, a distraction from "unremitting banality and inconceivable terror", but it can also serve to "normalize what is psychologically unbearable, thereby inuring us to it."[1] So, part of the appeal of sci-fi movies is so "one can participate in the fantasy of living through one's own death and more, the death of cities, the destruction of humanity itself." It's escapism that doubles as a way to navigate fear, but this and the normalization role meant that to Sontag, sci-fi was an "inadequate response" to major sociopolitical issues.

Similarly, to Andrew Tudor (*Monsters and Mad Scientists: A Cultural History of the Horror Movie*, 1989), these films glorified government; military, scientific, and bureaucratic elites were uniquely capable of providing for our defense. They "teach us not so much 'to stop worrying and love the bomb,' as 'to keep worrying and love the state.'" Peter Biskind, in *Seeing Is Believing: How Hollywood Taught Us to Stop Worrying and Love the Fifties* (2000), also regards them as representing "liberal-corporate sci-fi." Science might have created the crises in these stories, but it also often provided the solution as well.

As we'll see, this is very different from the sci-fi that followed, stories in which progress has led us to a place from which there might be no return. If 1950s sci-fi was about saving civilization, from the late 1960s onwards it would more often consider whether civilization itself was the problem from which we need to be saved (and probably won't be).

"I GUESS YOU GUYS AREN'T READY FOR THAT YET. BUT YOUR KIDS ARE GONNA LOVE IT."

Talking of traveling at accelerated speed into the future, in 1970, exactly halfway between the 1955 and 1985 we visit in *Back to the Future*, Alvin and Heidi Toffler published their famous book *Future Shock*. It became an international bestseller and launched their careers as probably the world's best-known prognosticators on coming trends.

Their argument was that we are experiencing "too much change in too short a period of time." This accelerated rate of change could be seen in the growth of cities, energy consumption, economic growth (the Tofflers concentrated on the postwar boom from 1948 to 1965), and of course technology, for example in how fast we can travel, from the steam train to space capsules.

To the Tofflers, this acceleration has created a new freedom from the kind of standardization and regimentation of previous eras. But it also creates anxiety: "Future shock is the dizzying disorientation brought on by the premature arrival of the future," they wrote. Indeed, "It may be the most important disease of tomorrow." They identified a variety of its symptoms, including "the spreading use of drugs, the rise of mysticism, the recurrent outbreaks of vandalism and undirected violence, the politics of nihilism and nostalgia, and the sick apathy of millions." Which all sounds rather prescient (we'll return to acceleration in later chapters).

Alvin Toffler also saw the political culture around him at the time as inherently backward-looking, regarding (somewhat patronizingly) both the new left counterculture and the new right as expressions of nostalgia (we'll get into both of them in later chapters): "The middle-aged, right-wing reversionist yearns for the simple, ordered society of the small town [whereas] . . . the youthful, left-wing reversionist dreams of reviving an even older social system . . . the exaggerated veneration of pre-technological societies and the exaggerated contempt for science and technology."[2]

For the Tofflers, then, right and left shared more than they would comfortably admit (as we'll see in later chapters, those who critique industrial capitalism are often dismissed as 'backwards-looking' and anti-technology). We'll come back to the 'simple, ordered society of the small town' in a moment. But as with any works of futurology (or sci-fi), the focus has often been on which of the Toffler's sociological predictions have or haven't come true. Again, that's not what really matters though. What does is that the Toffler's idea has stuck. Even fifty years later, commentators still refer to accelerating change and that people resist the future by standing against the 'inevitable' (this sense of future shock is integral to cyberpunk as well). But what if they, and the Tofflers, are wrong?

"WHERE ARE WE? WHEN ARE WE?"

Ironically, there's nothing new about future shock. As Marshall Berman has commented: "People who find themselves in the midst of this maelstrom are apt to feel that they are the first ones, and maybe the only ones, to be going through it; this feeling has engendered numerous nostalgic myths of pre-modern Paradise Lost. In fact, however, great and ever-increasing numbers of people have been going through it for close to five hundred years."[3]

But what about that nihilism and nostalgia, which does seem to capture something about our age? What explains that, if not future shock? As the critical futurist Richard Slaughter notes: "Conspicuously lacking were ways of understanding, and coming to grips with, other dysfunctional imbalances in culture. 'Change' is only one of them. Meaninglessness, lack of purpose, hyper-materialism, technological narcissism and spiritual hunger are a few of the others that might be encompassed within a wider view. But 'Future Shock' was silent upon them all."[4]

It's not just how much change of course, it's also what kind of change it is and who it benefits from it, especially if it only seems to create more insecurity in our lives. The Tofflers might have been right about some of the symptoms of future shock, but not its real causes.

In addition, despite its huge success, *Future Shock* was actually terrible timing. It really reflected the past, not the future. At the point the Tofflers were writing, economic growth was slowing. The 1970s would be better known as a period of 'stagflation' (lower growth combined with higher inflation), and despite what has often been claimed, ever since growth on average has been lower than during the postwar boom. Relatedly, the sense of Tomorrowland would also fade. To take just one example, those 1960s expectations that supersonic travel would become the norm have evaporated, and transport speeds (including in aviation) have largely stagnated as well. There would be no flying cars for us to be future shocked by, either.

In chapter 4, we'll consider the challenges to the idea of progress that were arising at the same time the Tofflers were promoting *Future Shock*. For the moment, let's note that arguably the only thing that has kept accelerating since then is environmental destruction, and the main thing we're going to be 'shocked' by in the future is civilizational collapse.

Are the McFlys future shocked? More likely, there's *too little* real change in their world, just the grinding disillusion of middle-class life under pressure and on the slide. It's the future done *to* people, continually reminding them of their powerlessness and lack of control. The McFlys are struggling to keep up. They don't seem tough enough for the acquisitive 1980s, let alone the future. Marty's father George, for example, is being mercilessly bullied by Biff, his supervisor, who gets George to write work reports for him.

It can hardly have helped the McFlys that in the 1980s they had the Reaganite new right telling them they had to work harder and be more entrepreneurial so that America, and they, could compete in a 'global market.' And then in the 1990s, *Future Shock*–style, Clintonite 'third way' centrists were telling them, in only slightly softer terms, that they had to be fitter, more flexible citizens by educating themselves for the twenty-first century. In both cases, politicians positioned themselves as representing the 'inevitable' future, and characterized their opponents as being hopelessly stuck in the past. But in neither case were they promising to direct the future for ordinary people, let alone granting the McFlys or anyone else the option of questioning just who this future is meant to be for.

Despite what they were told, if we caught up with the McFlys today, it's likely that even if they were better educated or more skilled, their income would have stayed flat or fallen. And they'd probably be pretty angry about it.

"NO, IT'S STILL HILL VALLEY, BUT I CAN'T IMAGINE HELL BEING MUCH WORSE"

We don't stay in the future for long. The 2015 Biff steals the time machine and travels back to 1955 to give his younger self a copy of a sports almanac with the results of major sporting events from 1950 to 2000 (effectively another book of the future), causing an even bigger crisis.

The alternate 1985 this creates (or as Doc calls it, '1985a') is the clearest example of a dystopia in the series. At the time, *Back to the Future: Part II* received a mixed reception, but more recently it's gained a cult and critical following. As Gale acknowledges, "[Audiences] were absolutely surprised by it. The whole 1985a stuff . . . we went places the audience was not ready to go. [But] That is some of my favorite stuff in the whole trilogy."

As a result of his 'uncanny' ability to predict the results of sports events, Biff is now rich and powerful, and he dominates Hill Valley. It's become a hellhole (the film's cinematographer Dean Cundey calls this the "Biff-Horrific period"). Biff has taken over most of the town's land, as well as ownership of the police department through his corporation BiffCo. He's also murdered Marty's father and has married (and is abusive to) Marty's mother.

In this Hell Valley, crime is rampant, and menacing bikers and armed teenage gangs roam around with impunity. The town's public schools have burned down, the library is closed, and the courthouse has been converted into Biff Tannen's Pleasure Paradise Casino and Hotel, which also proudly houses the Biff Tannen Museum. He's also built toxic waste disposal sites downtown and owns nuclear power plants. Gambling has been legalized. Downtown is thriving once again—but with strip clubs, peep shows, porn theaters, and brothels. It's greed and unregulated capitalism literally run riot.

Biff is now one of the wealthiest men in the country and uses his money to influence national politics. His exploitation of the sports almanac serves as a fair metaphor for class advantage: the super-rich can't lose, and their greed and lust for power turns everything into a dystopia (just not for them of course). In this alternate history, Biff helped Richard Nixon remain president until at least 1985 and the Vietnam War was still ongoing in 1983.

In a spin-off comic, in 1986 Biff runs for president himself, however he's killed in the past (thankfully). The real-life inspiration for the character was more successful. It's well-known that 1985a Biff (cruel, narcissistic, gaudy, wild hair) is modelled on Donald J. Trump, who by the 1980s had really begun to promote his brand of no limits American 'success.' Employing this symbol of selfishness, acquisitiveness, corruption, and moral decay is an effective counterpoint to the seemingly Reaganite (but actually not) sensibility of the first film. It also works thematically; the second and third films in the series skewer fragile masculinity and raging ego. Despite their notable differences, Marty and Biff share a thin-skinnedness and sense of entitlement, but only one of them learns to grow up.

It's so dystopian there's at least a couple of references to *Blade Runner*. The refinery stacks behind Biff's Pleasure Paradise resemble the hellish flaming towers from the beginning of Scott's film, and eagle-eyed viewers might spot the spinner (flying police car) parked in a driveway. Even without a sprawling metropolis or virtual worlds, it's rather cyberpunk: an all-powerful corporation dominating the populace, environmental collapse, public squalor and private affluence, co-opted police forces alongside lawlessness, and of course flying cars. But it's cyberpunk landed in small-town America, and despite the generally comedic tone, all the more unnerving for it. There's even a tank parked in Courthouse Square, presumably in case it's needed to maintain 'public order.'

But in many ways, 1985a's dystopia is just an extension of the decline and fall of Hill Valley that's been going on for decades. And Biff being Biff, he would always have yearned to dominate others and amass wealth. Absent the sports almanac, he probably would have found a way to do it eventually.

So, a bullying sociopath possessing a stolen advantage takes over small-town America and creates a dystopian nightmare. Ignore the exaggerated chaos, this is an 'ordinary dystopia', one that we can imagine coming about without some apocalyptic, external event. As we'll see in later chapters, this idea of a dystopia that comes about through ongoing, slow collapse is important, most obviously because it's coming true.

"I'VE ALWAYS DREAMED OF SEEING THE FUTURE, LOOKING BEYOND MY YEARS, SEEING THE PROGRESS OF MANKIND"

In many respects it's already here. Look at it this way: how many of the residents of 1985a Hill Valley wake up every day and think, 'Wow, I actually live in a dystopia, just like in the movies'? More likely, they think 'How am I going to get to work without being mugged today?' Some might even hang onto the idea that any day now Biff will make Hill Valley Great Again, just you wait and see.

Sci-fi stories can be instructive, but not when they distract from what's already happened. We can focus too much on the obvious tech-dystopia aspects at the cost of the ongoing collapse of security, prosperity, health, and happiness. As Gregory Claeys points out in *Dystopia: A Natural History* (2017), the word 'dystopia' as coined in the 1740s means 'an unhappy country.' The normality, even banality, in this definition is its significance.

So yes, we increasingly live in a world of high-tech mass surveillance, monolithic corporate power, environmental collapse, disinformation, and disorientation—all very cyberpunk. And that could well be our dark, neon-soaked future. But in more ordinary terms, there's the dystopia in our present: ever-growing inequalities, oligarchs who present themselves as 'men of the people', millions of children going hungry, healthcare beyond the reach of many, virulent white supremacy, racist and militarized policing, mass incarceration, inescapable debt, crumbling infrastructure, declining life expectancy, soaring mental ill health (especially among young people), drug addiction, rampant gun violence, political and social polarization, paralyzed political systems that are seemingly unable to do anything about any of it and don't seriously pretend they can, and the assumption (on left and right) that politicians are irrevocably corrupted—all of which only feeds the widespread feeling of powerlessness, frustration, and disengagement.

These are the social pathologies of collapse. Of course, many people have lived in this ordinary dystopia for years, some for all of their lives. But for many more, this creeping dystopia is more recent, including in once vibrant towns like Hill Valley, places that assumed they had a bright future to look forward to, if they had to think about it at all.

The paradox of course is that, as blogger Jason Kottke among others puts it, America is a rich country that feels like a poor country (as we'll discuss, it's been *made* that way). Accusations that disconnected metropolitan elites haven't recognized the depth of this collapse, this ordinary dystopia here and now, these low attention disasters, aren't wholly wrong. As Indi Samarajiva put it starkly in a blog post published in September 2020 ('I lived through collapse. America is already there'), explaining his experience living in Sri Lanka during its civil war: "As someone who's already experienced societal breakdown, here's the truth: America has already collapsed. What you're feeling is exactly how it feels. It's Saturday and you're thinking about food while the world is on fire. This is normal. This is life during collapse."[5]

"YOU'RE NOT INSTILLING ME WITH A LOT OF CONFIDENCE, DOC"

So, let's look at this 'normal.' Official rates (as measured by the U.S. Census Bureau) are that 38 million Americans (11.6 percent of the population) live in poverty—but this is based on the official poverty measure which has remained virtually unchanged since the mid-1960s. Even then, 30 percent of Americans (93.6 million people) live close to poverty, with incomes less than two times that of their 'poverty thresholds' (the poverty line, based on the size of a family and the ages of its members). The rates are even higher for children, Native Americans, Blacks, and Hispanics. According to the Organization for Economic Cooperation and Development, nearly a quarter (23 percent) of American workers labor in low-paying jobs (meaning earning less than two-thirds of median wages) – the highest proportion of any rich nation. One-half of renters spend one-third of their income on housing. Americans owe $1.78 trillion in federal and private student loan debt—and rising. Before the coronavirus pandemic struck, 27 million Americans lacked health insurance. Two-thirds of family bankruptcies are due to an inability to pay medical bills; one-third of all funds raised via GoFundMe go to healthcare costs.

As the engineer and futurist Vinay Gupta puts it: "It's not that 'collapse' is happening—what you call 'poverty' is collapse. . . . The thing about collapse is it's not until it happens to you or your immediate family or your very closest friends that you recognize it as collapse—up until then it's just economics. The collapse is already here, it's just not evenly distributed."[6]

Some readers will recognize the paraphrase in the last line. Cyberpunk author William Gibson once said that "The future is here, it's just not evenly distributed." More recently, he updated it, remarking that, "Much of the planet's human population, today, lives in conditions that many inhabitants of North America would regard as dystopian. Quite a few citizens of the United States live under conditions that many people would regard as dystopian. Dystopia is not very evenly distributed." Or as fellow sci-fi author Kim Stanley Robinson prefers to say, "The money is already here, it's just not evenly distributed."

Poverty, austerity, a lack of healthcare, a lack of access to food, poor physical and mental health, homelessness, vulnerability, loneliness, pollution . . . they've always been the dystopias in our midst, largely ignored because they happen mainly to poor people, or when they were noticed, regarded as dystopias they somehow brought upon themselves. 'All' that's happened is that the shadow has crept over more lives, more towns and cities, and more middle-class and hoped-for middle-class lives.

What's in the background of *Back to the Future* is what we know but struggle to acknowledge: in America, fewer people are middle-class now compared to the 1960s. As the financial and social commentator Charles Hugh Smith has set out, if we define 'middle-class' as a set of financial attributes that were widely considered an attainable norm in the 1960s, and if prosperity has been rising for the past fifty years, more households today should be able not just to meet these attributes but easily exceed them.[7] We're talking about healthcare costs, significant equity in a home, incomes that enable households to save regularly, retirement savings, a financial cushion if you lose your job, a reliable vehicle or two as required, some ownership of non-real estate and/or income-producing assets, and sufficient time for family, leisure, and self-improvement.

As Smith states, even if you do have some of these things, propping yourself up on a precarious façade of debt-funded consumption does not qualify you as middle-class. Another test: the middle-class accumulates capital that can be passed onto the next generation. This is why so many of those thinking of themselves as middle class feel poor; by the conventions of an earlier age, they are. And this is because costs have soared for things that were once affordable, while incomes by and large haven't.

Of course, this is also reflected in hugely widening inequality. The top 0.1 percent now own more of the American wealth than the bottom 80 percent. The three richest Americans hold more wealth than the bottom half of the population. But the key point is how this is a result of Reaganite neoliberal policies, and a massive reversal of previous trends. Between the Second

World War and 1980, US median household income more than doubled, but 70 percent of all increases in Americans' income went to the bottom 90 percent. Since 1980, nobody's income has doubled except for the richest one percent, and the incomes of 90 percent of Americans have gone up by only one-quarter. Since 1983, the share of national wealth owned by lower- and middle-income families has declined from 39 percent to 21 percent, while upper-income families have increased their share of national wealth from 60 to 79 percent. According to one study, 2.5 trillion dollars every year has been transferred upward to the richest Americans—an astonishing total of *47 trillion dollars* since 1975.[8] It's a Biff Tannen scale of looting.

Such accelerating inequality actually slows growth. It's also politically unsustainable. As Henry Wallich, a former governor of the US Federal Reserve, once admitted, "Growth is a substitute for equality of income. So long as there is growth there is hope, and that makes large income differentials tolerable." But when growth slows, as it has overall . . .

And if you're not middle class anymore, it also means there are far fewer members of the middle class than America likes to assume or admit. You are working class, which in American terms is tantamount to the end of the world. For a country effectively built on expectations of near-continual progress, it's the end of the popular future.

The baby boomers—Marty's parent's generation—are often accused of being the ones who looted the economy. But as we've seen, not all of them have benefitted, especially not compared to the super-rich, despite being told that they too would become wealthy.

"YOUR FUTURE IS WHATEVER YOU MAKE IT. SO MAKE IT A GOOD ONE."

In the third film, the battle against dystopia goes way back to Hill Valley 1885, at this point a frontier town at the beginning of its existence. 'Mad Dog' Tannen's gang effectively runs the town. Left unchecked, his descendants will rule over Hill Valley forever. Marty defeats Mad Dog, but only by rising above provocations to his fragile masculinity.

The past is put right and the future secured. Marty had to go back into the past to fix the broken American suburban dream. But of course, we don't have a time machine. The profound decline of American utopia and its turn to dystopia can't be fixed in the past, and probably not in the future either. *Back to the Future* is a critical fantasy. It's clever, entertaining, but also subtly scathing about the precariousness and ultimately illusory nature of this utopia and how easily it can collapse into a dystopia—right where you live.

As a time travel story, *Back to the Future* would seem to be based on the empowering notion that we influence our own destiny, that the choices we make, for example when we're young, affect how things turn out when we're older. Indeed, this is a repeated riff in the films. But it's not so simple. This also means the decisions made many years ago, not just by us but by many others, also determine our present and future. So it's not as if we can just decide our own destiny, as we're told by conservative ideologies of meritocracy in a free market society, that if we just work hard we stand some chance of securing middle-class material comfort. (It's too often forgotten that the term "meritocracy" comes from a subtle, somewhat dystopian satire, coined in 1958 by the sociologist Michael Young in his essay *The Rise of the Meritocracy*, in which a seemingly fair system is really a disguised mechanism for upper class rule.) And we need to dig into why.

It was supposed to be a parochial paradise. And for a time, it was, even though, as *Back to the Future* shows, it was never perfect. But in historical terms, the predominantly white, comfortable, middle-class small towns and suburbs *were* a kind of utopia, and the future looked even brighter. But dystopia was hidden in utopia, as it often is.

Back to the Future illustrates these hopes, but also the slide of this utopia. Lyon Estates, where the McFlys live in 1985, is a middle-class suburban development. But it represents the declining middle-class world, the loss of 1950s optimism; graffiti has even been sprayed on its entrance gates. It's years into the Reagan revolution but there's no sense of a family and their suburbia on the rise, no morning in America here (Reagan's sunny reelection campaign slogan that's cited in the nostalgic Cafe '80s that Marty visits in 2015).

When Marty first travels back to 1955, Lyon Estates hasn't yet been built. He hides the DeLorean behind a billboard selling not just homes but a prosperous, comfortable middle-class lifestyle: "Live In The Home Of Tomorrow . . . Today!" The billboard shows a house like the one the McFlys will live in thirty years later. But it's a façade.

Fast forward back to the future. When Marty returns to 1985, even though his family's middle-class status has seemingly been resuscitated, Hill Valley has not, and why would it? Its downtown appears just as fraying and fading as before, still slipping toward ruin.

In *Back to the Future: Part II*, Marty is thrilled to learn that his 2015 self lives in Hilldale: "Boy, I must have really made it." Hilldale was supposed to be the fancy new development a few rungs above Lyon Estates. But in 2015, it's no longer prime real estate either, and even though the McFlys live in a home full of futuristic gadgetry, they're as hard-pressed and harried as ever. No matter what happens, decline seems inescapable. And that's before we move on to the disaster of 1985a, by which time Lyon Estates is promoted as a 'Proud Member of The Biffco Family.'

So, the slow decline of middle-class small-town suburbia, from 1950s growth, confidence and pride in the public realm and new development, to a 1980s with a fraying at the edges of town and middle-class life on the brink, and an alternate 1985 hellscape that's been taken over by the rich and totally trashed. The suburbs have become slums, reflecting the endpoint of the suburbanization of poverty. The inner-city has come home to Hill Valley.

"IT'S A VERY INTERESTING STORY, FUTURE BOY, BUT THERE'S JUST ONE THING THAT DOESN'T MAKE SENSE"

The global financial crisis of 2007–2008 revealed the precarity of an economy built on Potemkin prosperity. Subprime loans were 'necessary' because so many people could never afford to buy a house, a function of the illusion that middle-class living was still within reach. Of course, when the interest rates for such loans rose, the illusion vanished. A crisis in inequality became a real estate crisis became a global economic crisis.

But there's another fault line running through the American suburbs. As engineer and land use planner Charles Marohn, the founder and president of the anti-sprawl nonprofit Strong Towns, argues, the American pattern of development is a "growth Ponzi scheme."[9] It creates the illusion of wealth, but now that's evaporating and with it the prosperity we've taken for granted.

The problem is that the infrastructure required was never financially sustainable. The massive growth of American suburbia was predicated on continued further expansion, on forever growth to pay off previous investments and obligations. But sprawling development patterns were never productive enough to sustain themselves. The pivot point came in the mid-1970s, when growth started to slow, and we know what happens to Ponzi schemes when that happens. The long boom would always turn to bust, which now it has. Again, 1950s utopia contained the seeds of its own, slow, dystopia.

No wonder Lyon Estates, and Hill Valley in general, were starting to look frayed and tatty. No wonder the McFlys' middle-class dreams were as well. If anything, *Back to the Future* becomes more prescient as time goes on, because it was fundamentally right about the past.

What happened to the future, and can we ever get back to it? Some commentators suggest we've became irrationally pessimistic (in part encouraged by dystopian depictions of the future), or as a result of over-regulation, a lack of entrepreneurship, irrational fears around technology, even by becoming accustomed to comfort.[10] We stopped striving, stopped believing. *We* let the future down.

But an alternative view is that the future was taken away from us, the result of a conscious, deliberate program of political and economic change, one that paradoxically presented itself as 'the future,' the only possible future, but which effectively stopped progress through soaring social inequality and declining public investment, and which has created a poorer, less productive, more fearful, more pessimistic future—our present.

As David Graeber noted, this puts neoliberalism (the policies pursued by Reagan and Thatcher) in a difficult position regarding its promises about technology and the future, namely its claims that capitalism fosters rapid scientific and technological growth, increases overall prosperity despite widening inequality, and so creates a more secure and democratic world for everyone: "It is clear that capitalism is not doing any of these things any longer. . . . [W]onders take the form of modest improvements (the latest iPhone!), rumors of inventions about to happen . . . , complex ways of juggling information and imagery, and still more complex platforms for filling out of forms."[11]

Back to the Future critiques this anti-future program and its social effects through its depiction of the urban environment and through people's lives, the McFlys' especially, and the greedy bullies like Biff who thrive as a result. Not bad for an eighties sci-fi comedy.

"WHOA, THIS IS HEAVY"

All we're seemingly offered by this future is insecurity and the need to work harder, be more efficient and productive, and learn to live with the pressure, pressure, pressure . . . Understandably then, a 'social depression' has settled over the 1985 McFlys and many other families—not future shock, but future malaise. To Charles Hugh Smith, social depression is the social cost of stagnation and inequality driven by financialization.[12] And as Johann Hari puts it, this also affects how people see the future—where once most middle-class and working-class families had some sense of security and could plan for the future, this has been closing as a direct result of deregulation and making it more difficult for workers to organize to protect their rights. Moreover, this used to afflict mostly people in the lowest-paying jobs; now precarity has been rising further and further up the chain.[13]

It's both an individual and collective psychological depression. Without safety, security, and a healthy sense of a positive possible future, of course people will feel ill. Futurelessness is a state of depression, the stuckness reflected in *Blade Runner* and all those recurring cyberpunk visions we discussed in the previous chapter.

If we struggle individually to imagine a healthy, happy future self, if there's no plausible possibility of progress, this fuels short-termism, a lack of impulse control, anxiety-led consumerism, addiction . . . personal dystopias which mirror our economy and society. To more people, no, it definitely doesn't feel like 'the future is what we make it.' And as we'll see in later chapters, our inability to envisage progress is not an accident, it's a *design*.

This is even reflected in our young, seemingly optimistic hero. It might be decades ago, but Marty exemplifies a very contemporary characteristic (we did say the films are prescient). Really, he's lost. He's not the yuppie-to-be of *Family Ties*, strolling confidently toward a successful future. He's what he accused of being; a bit of a waster, low-level depressed by his family's slide, frustrated by their passivity, and unsure of himself.

Marty isn't doing well in school. He's interested in things, certainly: science experiments (until they go horribly wrong), his rock 'n' roll band, a future with his girlfriend. But if the fortunes of his older brother (who works at Burger King) and sister (who doesn't seem to have a job) foretell his own, he's certainly not motivated by what happens after he graduates (if he ever does). And then there's his father of course, exploited at his white-collar job by his boss, and given his deep-seated fear of confrontation, probably by others as well.

Then Marty is thrown into a desperate mission, not trying to lift his family up, but save his own and his brother's and sister's existence—the dissolving family photo a neat metaphor in itself, as if they're disappearing from the middle class, which in American terms is like not existing at all.

Marty carries the weight of middle-class malaise on his shoulders. Dreams of rock music stardom might distract him temporarily from day-to-day despondency. Maybe he and his brother and sister have been hit by the early 1980s Reagan recession. Young people were especially hard hit by that downturn, as they have been by recessions ever since. Even in rich countries, they're at risk of ending up poorer than their parents. As noted, wages have been flat or falling for many people over the past couple of decades or so, but young, less-educated people have been especially impacted. The Great Recession following the global financial crisis, and now the pandemic recession, have led to persistently high levels of youth unemployment, preventing many young people from launching careers. And as we'll see in later chapters, because of economic and technological changes to come, the hope of a comfortable middle-class life is likely to stretch even further out of reach.

"I HAD A HORRIBLE NIGHTMARE. I DREAMED THAT I WENT . . . BACK IN TIME. IT WAS TERRIBLE."

For the McFlys and many others, the future has become a four-letter word. It's no surprise then that the possibility of somehow traveling back to the past could be seductive. Echoing the Tofflers, some liberal/left commentators see this solely in regressive terms, a desire to return to a social order designed for middle-class (white) men. It's true, as the sociologist Zygmunt Bauman puts it, that today's authoritarian nationalist politicians promise a 'retrotopia', an inverted utopianism which looks to restore a supposedly unspoiled, idealized past. There might indeed be some element of future shock here, but of social and cultural change rather than technological developments, of a tomorrow accelerating toward increasing diversity.

Most critically, to Max Haiven, author of *Revenge Capitalism* (2020*)*, it's about vengeance: "[A]uthoritarianism does not conspicuously promise a better, utopian future, and the oath to return to some past seem dubious, even to adherents. But it does promise a revenge on those who can be blamed for some perversion or deviation in the past."[14] But dismissing this politics as mere nostalgia (or even revenge) neglects the understandable desire to return to security, comfort, leisure, optimism—to resurrect a desirable future. For many voters who have experienced deteriorating economic conditions since at least the crisis of 2008 and often for much longer, the promise of turning back the clock by, they're told, deporting immigrants and reclaiming lost sovereignty from international 'liberal' elites is compelling. It offers hope in a largely hopeless political climate.

Some liberal elites have been bemused by how authoritarian nationalists have described the state of things. They mocked Donald Trump's 2017 dystopian 'American carnage' inauguration address which described a broken country of abandoned factories, rampant crime, and failed schooling, just as they were surprised by his 'shock' 2016 election victory (the same year that Alvin Toffler died), after a campaign propelled by widespread anger and frustration among 'forgotten men and women' in 'left behind' places.

Often, these men and women personally know the cost of the future. To take just one example, according to a 2018 paper by Oxford University researchers, there was a correlation between automation and how people voted in the 2016 presidential election. If factory robots had instead barely increased in the years leading up to the election, Hillary Clinton would have won Michigan, Pennsylvania, and Wisconsin, and so the presidency.[15]

More jobs will be lost in low-paying transportation, retail, fast food, and call center sectors. To the debate about whether support for politicians like Trump represents white backlash or economic conditions, it's surely both:

economic precarity and prevailing decline makes people vulnerable to the racist blaming strategies of authoritarians.

Similarly, perhaps some of the residents of Hill Valley will have welcomed Biff's town takeover. He's a businessman after all, perhaps he can turn this place around, maybe only he can fix it (and everyone else seems to have failed) – ignoring of course that it's exactly what he represents that caused the decline in the first place, and even if, as we've seen, this political promise is about as realistic as a time travel machine hitched to a DeLorean.

"WHATEVER YOU'VE GOT TO TELL ME, I'LL FIND OUT THROUGH THE NATURAL COURSE OF TIME"

At the end of *Back to the Future: Part I*, Biff the bully gets knocked out and loses his power. But as we know from the rest of the story, that's not the end of Biffism, not at all.

Even though Marxist political economists are often best at recognizing the crises at the roots of right-wing authoritarian populist politics, they largely haven't recognized it as early collapse politics, as a response to the dystopias to come. As we'll see later, recognizing this prompts us to understand how conservative endgame politics will roll out—a preview of the future in the present.

It's not just a crisis of suburbia of course, it's much wider and deeper than that, as we'll discuss. As Sam Kriss argues, capital effectively cannibalizes its own future by spending resources it doesn't have yet, and so: "We're already changing the future: like the photo of Marty McFly and his siblings, our stock of time is fading in front of us. . . . [W]hat does the demand for a future even mean when . . . the future is now something that's *already happened*?"[16]

There's no real way back to the future—not to 1950s optimism (with or without its racism and sexism), nor to a future-future with a new sense of optimism. We're out of time.

NOTES

1. Susan Sontag, "The Imagination of Disaster," *Commentary*, October, 1965.

2. Alvin Toffler, *Future Shock* (New York: Random House, 1970), 358.

3. Marshall Berman, *All That Is Solid Melts into Air: The Experience of Modernity* (New York: Verso, 1982), 15–16.

4. Richard Slaughter, "Future Shock Re-assessed," Metafuture.org, undated, 2002. www.metafuture.org/articlesbycolleagues/RichardSlaughter/futureshock.htm

5. Indi Samarajiva, "I Lived Through Collapse. America Is Already There," *Gen*, September 26, 2020. https://gen.medium.com/i-lived-through-collapse-america-is-already-there-ba1e4b54c5fc

6. Emerge, "Vinay Gupta—Waking Up in the Monster Factory (pt. 1)," *Emerge: Making Sense of What's Next* podcast, April 1, 2019. https://anchor.fm/emerge/episodes/Vinay-Gupta---Waking-Up-in-the-Monster-Factory-pt--1-e3ikvk

7. Charles Hugh Smith, "Are You Really Middle Class?," *The American Conservative*, September 23, 2015. www.theamericanconservative.com/articles/are-you-really-middle-class/.

8. Carter C. Price and Kathryn A. Edwards, *Trends in Income From 1975 to 2018* (Santa Monica, CA: RAND Corporation, 2020). www.rand.org/pubs/working_papers/WRA516-1.html.

9. Charles Marohn, "The Growth Ponzi Scheme, Part 4," *Strong Towns*, June 14, 2011. www.strongtowns.org/journal/2011/6/16/the-growth-ponzi-scheme-part-4.html.

10. According to commentators such as J. Storrs Hall, among others, particularly culpable is government over-regulation stemming from a turn against technological progress and economic growth in the late 1960s and early 1970s, see J. Storrs Hall, *Where Is My Flying Car?* (San Francisco: Stripe Press, 2021). We discuss this 'turn' more in chapter 4.

11. David Graeber, "Of Flying Cars and the Declining Rate of Profit," *The Baffler*, 19, March, 2012. https://thebaffler.com/salvos/of-flying-cars-and-the-declining-rate-of-profit.

12. Charles Hugh Smith, "Our Social Depression," Of Two Minds.com, undated. www.oftwominds.com/blogmay15/social-depression5-15.html.

13. Johann Hari, *Lost Connections: Why You're Depressed and How to Find Hope* (New York: Bloomsbury, 2018), 141, 245.

14. Max Haiven, "Capital's Vengeful Utopia: Unpayable Debts from Above and Below," *L'Internationale Online*, May 11, 2020. www.internationaleonline.org/research/real_democracy/131_capitals_vengeful_utopia_unpayable_debts_from_above_and_below/

15. Carl Benedikt Frey, Thor Berger, and Chinchih Chen, "Political Machinery: Did Robots Swing the 2016 US Presidential Election?," paper, 2018. www.oxfordmartin.ox.ac.uk/downloads/academic/Political%20Machinery_171008_CF5.pdf.

16. Sam Kriss, "Abandon the Future," *Jacobin*, November 21, 2015. www.jacobinmag.com/2015/11/back-to-the-future-day-marty-mcfly-sam-kriss.

Chapter 3

"Oh, Wouldn't It Be Great If I Was Crazy? Then The World Would Be Okay"

No-One Believes the Plague-Ridden Future of 12 Monkeys *Could Actually Happen*

With *12 Monkeys* (1995), Terry Gilliam directed one of the definitive dystopias of the 1990s, after he did the same with *Brazil* in the 1980s. The film's 'recycled future' production design echoes its story, how the future is made up of the past. In a similar way, because we've all seen so many depictions of the future (and because we already know what might happen), mentally we're also effectively living both in our time and the future. That's what recurring depictions of dystopia really are, nightmare-memories that have haunted us for a long time. Contrary to techno-optimists, the dark future is not a self-fulfilling prophecy we could avoid if only we were more positive; it's a future reality we've already foreseen.

"I'M HERE ABOUT SOME MONKEYS"

Close-up on a boy with blue-grey eyes. He watches in terror and confusion as a man running through an airport concourse is shot in the back. A blond woman weeps by the man's side as he lies on the ground. The boy seems particularly affected by the distress and tenderness of the woman as the man dies in her arms.

In 2035, James Cole (Bruce Willis) is haunted by this nightmare, or perhaps it's a memory. He wakes up, but in a dystopian future. A virus has decimated the world, sending the remaining humans underground to live in a gloomy nightmare of control, surveillance, and paranoia—seemingly a world under permanent emergency measures.

Even worse, Cole is a prisoner. He's been 'volunteered' to go above ground to collect data. "I didn't volunteer," Cole insists, as prison guards forcefully dress him in a protective plastic suit. Above ground, the frozen city of Philadelphia is wrecked and abandoned. The surface of the planet has been reclaimed by nature. Wild animals, even (incongruously) bears and lions, roam deserted plazas and department stores.

If *Back to the Future*, at least at first, appears to offer the possibility of changing the future through the past, the next dystopian time travel film suggests we too are prisoners of time. With *12 Monkeys*, Terry Gilliam directed one of the definitive dystopias of the 1990s, after he did the same with *Brazil* in the 1980s (we'll discuss *Brazil* in chapter 8). The story starts in a plague-ridden future. And even though we travel back and forth through time, it's all dystopia really, spanning the trenches of the First World War, a mental institution, a darkly paranoid present, and sliding towards a predetermined doom.

The filmmakers searched for real-world locations that represented dystopias in our midst. As cinemaphotographer Roger Pratt says, "The kinds of places we ended up shooting at were disused theaters and ugly sections of town—abandoned places that had faded and fallen into disrepair." And the story as a whole is about whether the dystopian future might possibly be improved upon—whether there's any chance of a future *after* the future.

Until then, as in *Blade Runner*, it's a frozen future. Emphasizing this, the production design of *12 Monkeys*' underground world only uses pre-outbreak era technology. This reflects Gilliam's preferred junk heap aesthetic. As film critic Roger Ebert put it, "It has the same view of the near future as a grunge pit [as] Gilliam's own *Brazil*." Nothing in the film is sleek or foolproof, including its time-travel machine, which regularly malfunctions.

Neatly, this recycled future echoes the film's story, how the future is made up of the past, rather than being shiny brand new. Indeed, dystopian futures and premonitions of disaster have become so familiar they feel like recurring nightmares, even memories. We didn't believe ourselves, didn't believe that these premonitions could become real. And then they did.

"I THINK I'VE SEEN THIS MOVIE BEFORE"

Cole is sent into the past, our present, to locate the source of the virus, specifically before it mutated after its release, in order to help a group of scientists from his time develop a vaccine for the remaining survivors in the mid-2030s.

As Gilliam puts it, "It's about remembering and using the past to try and save the future."

So, as in *Back to the Future* ten years earlier, a traveler comes back from the future. And as in that film and its 1950s precursors, there's the theme of science going out of control. But unlike in most time travel stories, there's no apparent hope or intention of averting disaster (and the future dystopia is in no way as cyberpunk cool as that in *Blade Runner*).

We're familiar with the head-scratching paradoxes that time travel stories can throw up. But *12 Monkeys* appears to accord with the Novikov self-consistency principle (developed by the Russian physicist Igor Dmitriyevich Novikov in the mid-1980s) that 'whatever happened, happened.' No matter what anyone does, you can't create a paradox, because history is history. For example, if Cole did manage to stop the virus before it spread, there would be no dystopian future for him to be sent back from.[1]

Neither do the future scientists seem to want to. One of them later warns Cole not to become addicted to what he scornfully calls "that dying world" (our present). So, the crisis can't or won't be prevented, and adaptation means huddling underground. There's no way to unmake the past, and only the slim possibility of a kind of too late (for most) miserable mitigation.

Fittingly, *12 Monkeys* was itself inspired by Chris Marker's 1962 experimental short film *La Jetée* (the jetty or pier). *La Jetée* is a 'photo-roman', mostly made up of still photographs apart from a fleeting final shot. Film producer Robert Kosberg was a fan and persuaded Universal Studios to acquire the rights to remake it as a full-length feature. David Webb Peoples and Janet Peoples were hired to write the script. David Peoples had previously written *Blade Runner*, another dirty dystopia, and as discussed, also a seemingly resigned future.

As in *12 Monkeys*, in *La Jetée* the main character is haunted by a death in the past. Similarly, he's never sure if the image is dream or reality, though it's hinted at in the opening shot that he may already have witnessed the end of the story. (In *12 Monkeys*, Cole's dream-premonition changes subtly depending on what he's learning in the past.)

Marker's film is a critique of postwar France. Born Christian François Bouche-Villeneuve, Marker was a French writer, photographer, documentary maker, multimedia artist, and film essayist (and Marxist). In *La Jetée*, a survivor of a third world war is forced to participate in grueling experiments that send him back to 1960s Paris and then the future, to find a power source to save (or at least rebuild in some way) his own time. He's chosen partly because of the strong images he has of the past, particularly the man's murder.

La Jetée alludes to concentration camps, imperial atrocities, dehumanizing modernization (symbolized in the film by the new pier at Orly airport), and the threat of nuclear annihilation. Its apocalypse reflects that era's

Cold War anxiety. *12 Monkeys*' apocalypse is grounded in our anxieties of late-twentieth-century annihilation. Some reviewers reference the AIDS epidemic, but clearly the likelihood of a pandemic directly informed the filmmakers (for example, the time machine in the film is based on the isolation chambers used in outbreaks like the Ebola virus). More broadly, as we'll discuss, we can also say it's influenced by the increasing complexity and interdependency of contemporary society and our worrying vulnerability to a civilization-ending event.

"HE THINKS HE COMES FROM THE FUTURE"

Even for a time travel story, Gilliam's film particularly seems to reject linear time. It's set in 2035, 1990, 1914 (briefly), and 1996, but in each moment it feels both like now and not now. Unlike say, *Back to the Future* (as that film's title suggests), we never feel anchored in one time around which the rest of the story revolves, one place we can hope (or would want) to return to.

In a similar way, in our world, because we've all seen so many depictions of the future (and because we already know what might happen), mentally we're also effectively living both in our time and the future as well. As Abraham Riesman suggests: "[As] in *12 Monkeys*, it feels increasingly as though our minds are situated in the years to come, as though we're already looking back on what's happening to us right now from the vantage point of the coming calamities. . . . More and more of us are united by a feeling like anything good is living on borrowed time."[2]

Cole arrives in the past, but in 1990, not in 1996 as intended. He's arrested. On the diagnosis of psychologist Dr. Kathryn Railly (Madeleine Stowe), who takes an interest in his case, he's incarcerated in a mental hospital. Cole is interviewed by a panel of doctors where he tries to explain that the outbreak has already happened and that nobody can change it. He's deemed to be mentally ill. He, and his message, arrive too soon to be heard.

David and Janet Peoples were initially reluctant to take on the project: "We couldn't see the point. [*La Jetée* is] a masterpiece and we didn't see that there was any way to translate [it]." But watching it again, they started to see the possibility of a different take: "How would we react to people who showed up and said, 'Oh I've just popped up from the future' and in turn how would that person deal with our reaction?" As Janet explains, "We were very interested in asking questions like 'Is this man mad? And how about the prophets of the past, were they mad? Were they true prophets?'" (Their interest was also piqued when they came across an article about scientists digging into the permafrost to collect samples from the 1918 pandemic in the hope of preventing a future outbreak.)

Gilliam's films often address sanity, madness, reality, and fantasy. As he says, "The story is disconcerting. It deals with time, madness, and a perception of what the world is or isn't. It is a study of madness and dreams, of death and rebirth, set in a world coming apart." But in *12 Monkeys*, despite the director's trademark use of wide-angled lenses and Dutch (tilted) angles, we never really wonder whether Cole might be crazy. We've already seen his dark future, and perhaps because we've watched so many dystopian movies, we trust that his future is real (even though, as we'll see, later even Cole will doubt his own story).

Like the previous films we've discussed, *12 Monkeys* seems increasingly prescient as we approach the future. Cole desperately tries to convince people that they're in the past and that the true present is the postapocalyptic dystopia from which he's come. Despite the film's commercial success, to 1990s audiences this might have been received just as a neat idea, living as they were after the declaration of the 'End of History' (we'll discuss that idea in a later chapter), an era which even at the time felt like an endless present of growth, consumption, and pop culture triviality. It's not a neat idea now, as we increasingly experience the future we were warned about.

As Cole says later in the story, watching a film that feels both familiar and unfamiliar to him: "Like the past, the movie never changes. It can't change, but every time you see it, it seems different, because you're different. You see different things."

"LOVE THE AIR HERE. VERY FRESH. NO GERMS."

Twenty-five years after *12 Monkeys*, we experienced a real pandemic (perhaps distastefully, the film received a 2020 limited DVD re-release with a 'Free 12 Monkeys pandemic face mask'). As destructive as it was, the coronavirus wasn't as devasting as the pandemic featured (but not actually seen) in *12 Monkeys*, or in most fictional virus-based apocalypses and postapocalypses. But in many ways, as we'll discuss, we made the real pandemic far more destructive than it needed to be.

Susanne Kord, who's written an excellent analysis of *12 Monkeys*, notes that it's unusual among outbreak movies in that it encourages us to focus less on the ending (will humanity find a cure?), and more on life under a pandemic (what will humans do to survive?).[3] Indeed, as we'll see, how the story ends is ambiguous, but it hints at something horrifying.

Kord has also noted how the film echoes the coronavirus pandemic.[4] *12 Monkeys* states that we brought it on ourselves (we'll see how a bit later). We can further note that, dramatically, the film might point the finger at a single individual, but thematically it identifies a broader civilizational problem, and a nihilism about our ability to survive.

In the mental hospital, Cole meets Jeffrey Goines (Brad Pitt), a patient with strong environmental and anti-corporate views. Jeffrey mouths off about capitalism and the 'evils' of modern science: "Buy a lot of stuff, you're a good citizen. But if you don't buy a lot of stuff, if you don't, what are you then, I ask you? What? Mentally ill." Specifically, Jeffrey rails against the experiments conducted by his father, a prominent scientist whose virological tests on monkeys particularly drive Jeffrey into a frenzy. . . . But there's an important difference between Cole and Goines. Cole is a prisoner, although we never learn what crime he's supposedly committed. Goines is from a wealthy family. As he threatens the staff of the hospital, "My father will get me transferred. . . . Wait till you morons find out who I am! My father's going to be really upset." He also uses his status to claim he's not crazy and has the means to survive the coming apocalypse.

As Kord suggests, humanity in *12 Monkeys* stands divided in face of the apocalypse, split into a ruling class (including the scientists searching for a vaccine), and what appears to be the caged, exploited remainder of humanity who are subjected to scientific experiments. As obviously abhorrent as this is in fiction, the coronavirus, or rather our response to it, has again demonstrated how 'we' regard some lives as being worth more or less than others. Those most vulnerable to the coronavirus were those with less freedom of movement: prisoners, people in care homes, the often low paid 'essential workers' unable to work from home, poorer people and minorities, all the while wealthier people could shelter comfortably in place or even escape elsewhere.

Cole, when he's sent on missions to the present plague-ridden surface or the pre-plague past, is expendable. Tomorrow's rulers are happy for him to be exposed to environmental hazards, just as today we're seemingly comfortable for poorer communities to be exposed to pollution, toxic air, and the consequences of climate change. As we'll note in chapter 7, we've been misled by many movie apocalypses: future collapse won't affect everyone equally. Why would we expect it to, when we look around our own world?

"YOU WON'T THINK I'M CRAZY WHEN PEOPLE START DYING NEXT MONTH"

As Kord suggests, *12 Monkeys* also dramatizes the cost of human captivity and isolation. Its human cages, stifling subterranean facilities, and full-body protective gear might be much more extreme than most people experienced in the coronavirus. But all of us can now relate to the idea of the outside world being a potentially toxic space, not to mention dealing with death, grief, loneliness, the loss of our sense of time, and lost futures.

Cole in particular is repeatedly trapped: in a future prison, in a mental institution, and in the knowledge of what will happen from his own nightmare/memory. Just as we might be by the future that's already been made in the past.

Perhaps most significantly, to Kord, in contrast to most viral outbreak films, *12 Monkeys* presciently dismisses our hopes of a return to normal after the coronavirus pandemic or any other such destabilizing event: "[The film] refuses to push the re-set button, rejecting the comforting idea that you can always start over."[5] Nevertheless, the desire for normality was strongly felt during the coronavirus outbreak. Some conservatives railed against 'illiberal restrictions' and advocated 'herd immunity' that in reality would mean 'sweeping away' the most vulnerable and those without health insurance. As Kord asks, "How far away are we now from openly hoping that in the post-virus world, the elderly, the poor, the homeless and the chronically ill will be gone, saving our government big bucks on health and social care?" Not all that far, it seems.

After an attempted escape, Cole is locked in a cell. But moments later he disappears, waking up back in 2035. The scientists play Cole a distorted voicemail message that suggests that a fanatical group called the Army of the Twelve Monkeys was responsible for releasing the virus. They also show him pictures of people suspected of being involved, including Jeffrey Goines. . . . That makes sense: environmentalists hate humanity, right?

It's November 1996, six years since Railly's encounter with Cole, and it's changed her. She's been researching people who claimed to be from the future but were disbelieved, and she's published a book on the subject titled *The Doomsday Syndrome*. Now she's giving a lecture based on her book to a group of scientists: "Cassandra in Greek legend, you recall, was condemned to know the future but to be disbelieved when she foretold it. Hence the agony of foreknowledge combined with the impotence to do anything about it."

Railly discusses the 'Cassandra complex', the psychological phenomenon in which a person's predictions of crisis are ignored. The term is derived from Greek mythology. Princess Cassandra was the daughter of Priam, the king of Troy. Apollo, enchanted by her beauty, gifted her the ability to foresee the future, but when she rebuffed his advances, he took revenge by ensuring that no-one would believe her. She was fated to know the future, including the fall of Troy and her own death, but unable to change it.

Railly's book discusses numerous doomsday prophets. As an audience, we realize these might be other travelers from the future, warning about the fate of humanity. Of course, at this point Railly doesn't believe that Cole actually could know the future. But after the lecture, another scientist, Dr. Peters, challenges her by suggesting that such people carry rational warnings, while our destruction of the environment represents the real lunacy. Indeed, scientists and environmentalists have been sounding the alarm bell for some time (we'll get into that in the next chapter). As have novelists, writers, filmmakers, and artists. Unfortunately, few of us listened.

"I DON'T BELIEVE IN GERMS. GERMS IS JUST A PLOT THEY MADE UP SO THEY CAN SELL YOU DISINFECTANTS AND SOAPS."

In 2035, the scientists offer Cole a second chance to complete his mission and send him back in time again. Cole accidentally arrives at a battlefield during the First World War. He's shot and wounded in the leg, and is suddenly transported to 1996.

There, Cole kidnaps Railly and forces her to take him to Philadelphia. They learn that Jeffrey is the founder of the Army of the Twelve Monkeys. When they confront him, he denies any involvement and says that, in 1990, it was Cole who originated the idea of wiping out humanity with a virus stolen from Jeffrey's virologist father, Dr. Leland Goines. Just as he's about to be caught by the police, Cole is transported back to 2035, and then to 1996. When he finds Railly again, he tells her he now believes his future is just a delusion.

We can empathize. Recognizing collapse can make us wonder whether we're deluded. After all, everyone else just carries on as if civilization will last forever, that the future isn't going to happen, at least in their own lifetimes. Like Cole, we might end up willing ourselves to believe that we're imagining it. It would be so much easier that way.

However, based on his predictions and photographic evidence from the past, Railly now believes that Cole has been telling the truth. She's been drawn into the Cassandra complex herself, and is convinced that the Army of Twelve Monkeys does indeed pose a threat. But feeling they can't stop fate, Cole and Railly decide to depart for the Florida Keys before the onset of the plague—to live, just for a time, before the inevitable end.

Another way of deflecting doom is to believe it's all a hoax. Often, we regard conspiracy theories as a rejection of humdrum reality, a reluctance to recognize that some things just happen, that there's no grand plan behind events. Instead, it must be because someone powerful has planned them—or maybe they aren't really happening at all.

We've certainly heard that, in the coronavirus pandemic and with other crises such as climate change. Indeed, the future depicted in *12 Monkeys* is a conspiracy theorist's heaven (or hell): a society locked down under permanent emergency measures, a giant medical prison controlled by an elite cabal of experts, regular citizens branded with barcodes, caged like lab rats, and forced to put themselves forward for lethal experiments.

The film was released just before the rise of the anti-vaccine movement, but it reflects some of their suspicions about the supposedly insidious designs of the medical establishment (at one point, in the belief they contain tracking devices, Cole even removes his teeth to stop the future authorities from tracing him; weirdly, it seems to work, at least for a while).

It's doubly difficult when the authorities themselves promote conspiracy theories to deflect blame from their own incompetence, or worse. (As president, Donald Trump flirted with the idea that the coronavirus was deliberately released by a lab in China, and that sensible measures like wearing a mask were unnecessary restrictions on personal liberty.)

But of course, the virus is real. And accepting this is the start of engagement with the much more frightening notion that the real explanations for crises are deeply rooted in contemporary civilization, in the ways that we all live and have come to trust as sane and rational. In our frantic search for the 'real culprits', our susceptibility to insanely complicated conspiracy theories distracts us from what's been hiding in plain sight all along.

"YOU'RE HERE BECAUSE OF THE SYSTEM"

In futures studies, 'backcasting' is a way of considering how we might achieve a desirable future. We start by defining the future we want and then work backwards to identify the actions that will help us reach that future. Much science fiction (especially nearer-future sci-fi) is a form of backcasting, or as the utopian-leaning novelist Kim Stanley Robinson puts it, a kind of modeling exercise, a 'future history' making.

But echoing *12 Monkeys*, what if we 'doomcast' instead, travelling backwards from a recurring dystopian vision and identifying the actions (or non-actions) that could lead us there? (We're not trying to change the past, just understand it.) As we discussed in chapter 1, contrary to some critics this is where the radicalness of some dystopian stories lies, as a way into considering the deeper causes of collapse. As in *12 Monkeys*, could this knowledge help us to save the future, or if disaster is unavoidable, at least some part of it?

In chapter 8, we'll discuss the rise of survivalism based on the belief in impending social, economic, and political collapse. This grew in the 1990s, and especially in the 2000s after the 9/11 terror attacks. For the moment, we can note that the typical survivalist story, whether supposedly factual (as anticipated in survival guides) or fictional (in the fast-growing genre of survivalist novels) is based on a single devastating event—a power grid failure, solar flare, breakdown in satellite communications, or yes, a (often deliberately engineered) pandemic.

Understandably, for dramatic reasons and audience comprehension, fiction tends towards the big single event. But as we'll discuss, starting in the next chapter, the reality is likely to be not a cataclysmic event, rather a systemic, interrelated sets of crises which slowly degrade our ability to secure our societies in their present form. And after all, events alone aren't necessarily catastrophic; it's how resilient our societies are, and how we respond to them, that's critical.

But it's more than that. We noted in the previous chapter the growing sense that America is a rich country that feels like a poor one. As numerous commentators also suggested during the coronavirus pandemic, in some respects it also increasingly resembles a failed state. Not only are *we* responsible for turning crises into disasters (and ultimately dystopias), but existing problems we've failed to solve can exacerbate the crises that arise.

There were numerous comment pieces at the beginning of the pandemic that it could act as a social leveler, since 'microbes don't respect socio-economic class.' But in a way, they do. This goes back to the point made in the previous chapter: many people already live in dystopia, or encroaching dystopia. Their existing conditions represent crisis right now, we just don't recognize them as such. In the case of the United States (and other countries long dominated by so-called new right or neoliberal ideology such as the UK), high levels of economic inequality, and in health and well-being, acted as a breeding ground for the virus.

Moreover, inequality was disastrous in exacerbating the crisis. Poverty, precarious employment, a lack of social safety net and universal access to quality healthcare, hollowed-out public infrastructure, and more generally growing levels of distrust in government and expertise, all contributed massively to the coronavirus spreading more quickly and widely, and many more people suffering and dying, than could have been the case. In other words, there was a preexisting condition that made the virus so deadly. As Dr. Abdul El-Sayed, epidemiologist and author of *Healing Politics: A Doctor's Journey into the Heart of our Political Epidemic* (2020) puts it, it was an epidemic of insecurity.[6]

The regime in *12 Monkeys* might seem like a world away from even the often-disastrous US federal government response to the coronavirus, and in literal terms it is. But you don't end up with *12 Monkeys*' regime because of just one pandemic, you get there (or edge ever closer to it) because your society is already failing, already starkly unequal, divided, threadbare, and paranoid, as *12 Monkeys* also rather suggests the United States was and is.

In many ways then, the coronavirus 'merely' accelerated already apparent trajectories of contemporary capitalism: rapidly gentrifying cities, automation and the gig economy displacing workers into precarious forms of employment, young and already marginalized people losing their jobs, seemingly ever-rising costs of living, and lack of access to affordable healthcare, education, and daycare for children. And then a flood of very sick people attempting to access a broken, hugely unequal healthcare system that was unprepared to handle anything like a pandemic. Apparently, the United States had only 45,000 intensive care beds at the start of the pandemic (South Koreans for example have more than three times more beds available per thousand people), and local and state health departments have a quarter less staff today than they did before the global financial crisis.[7]

More broadly, the threat the virus posed was facilitated by the highly unequal world that industrial capitalism has created. The destruction of environmental habitats, corporate farming, rapid urbanization, and weakened social safety nets create the conditions for viral outbreaks. Climate change and ecological breakdown is a further petri dish for disease, creating the real possibility of ancient bacteria being released from melting ice. Pollution, especially affecting poorer communities, also weakens people's immune systems. As Mike Davis put it, "Coronavirus shows that capitalist globalization is biologically unsustainable."[8]

It's the combination of a particular model of globalization, interconnectedness, and economic contagion, austerity in the public sector, and a lack of planning and preparation in the private sector because it's not profitable, that adds to all this damage. Just as the coronavirus first showed up among the most vulnerable—the old, the sick, the working poor, and racially oppressed—it has also taken aim at the weakest parts of global, national, and local systems. From a broken healthcare system, racial injustice, economic inequality, brittle supply chains, climate threats, political corruption, and a frayed social fabric, the virus revealed a complex set of system failures. It's a premonition of collapse. In other ways, it distracted from the collapse already going on all around us.

As to how we reached this point, even before the pandemic, well, we'll get to that too.

"THEY FIND US THROUGH OUR TEETH!"

Another thing that was previewed in the coronavirus, which is also dramatized in *12 Monkeys* and many other dystopias, is how crisis can lead to greater surveillance. Clearly, protecting public health can require the effective use of technologies. But 'emergency measures' can easily become permanent, and normalize how we expect states and corporations to track us and know ever more about our lives. Moreover, there's how those bodies think about us during such crises. In a 2008 report on pandemic responses, a group of historians and medical ethicists assembled by the American Civil Liberties Union criticized the tendency for government to address public health problems using mindsets more appropriate to tracking criminals. Paradoxically, such tactics can also make fighting diseases harder by increasing distrust between government and citizens. As the ACLU puts it, "People, rather than the disease, become the enemy."

We can debate the extent to which big technology companies represent a new mode of 'surveillance capitalism', and whether it's sensible for us to trade apparent convenience for confidentiality in our everyday lives. But when we consider the context of coming crises, without lapsing into conspiratorial paranoia, the real danger is how their knowledge and ability to track us could easily slip into digital (or physical) prisons and other forms of control—all justified as 'necessary' of course as crises escalate.

Hoping *they* aren't being tracked, Cole and Railly hurry to the airport. On the way, they learn that the Army of the Twelve Monkeys is not responsible for the virus, but they still don't know who is, or where the actual conspiracy lies. The group's only act has been to release animals from the zoo and imprison Jeffrey Goines' virologist father in a cage. At the airport, Cole leaves a last message telling the scientists that the Army of the Twelve Monkeys is not to blame but that he won't be returning to the future. He has become addicted to this dying world after all.

In our pandemic, we're still chasing explanations about where the coronavirus came from. We may never know for certain, but we have some leads. Pandemics have been with us throughout history. But emerging infectious diseases are increasing significantly in direct correlation with human population growth and our ecological footprint. Deforestation, industrial-scale animal farming, road building, and sprawling development increase the contact between humans and virus-carrying animals. We've colonized the planet and now we're encountering viruses we've never experienced before. Paradoxically, the more we encroach on wild habitats, the more the safe space for human habitation shrinks.

Terrifyingly, between 10,000 and 600,000 mammal viruses are estimated to have the potential to spread to human populations, the vast majority of which are currently circulating in wildlife, undescribed and undetected by disease outbreak surveillance.[9] Their transmission is made more likely due to how industrial expansion is driving climate and land use changes. Climate change mitigation is unlikely to reduce this risk, even if we do manage to stay under two degrees Celsius of global warming, which is now unlikely. As Professor Colin J. Carlson of Georgetown University's Center for Global Health Science Security says, "It's coming no matter what we do now."

12 Monkeys correctly references rampant consumerism and unbridled capitalism. Behind these, the systems of finance, trade, communication, and transportation that underpin industrial production, that have previously provided for the health, well-being, and comfort of populations, are now becoming our sources of greatest threat. As we'll see in later chapters, we're entering one immense, *predicted*, civilizational crisis. The coronavirus pandemic is just one 'sign' that we've entered into an increasingly volatile new phase of chronic, civilizational instability, not from one crisis but the interaction of many crises—from climate change to food system failures, financial crises to civil unrest. It's the escalating synergy between these crises which is pointing to collapse and chaos.

As Gilliam says, "My biggest concern is who is going to be managing the cull of humanity. . . . We consume everything, and I don't see how we can keep taking everything we seem to think we need from the world, from nature, and the whole thing not fall apart."

You're not delusional, it really is happening.

"HOW CAN I SAVE YOU? THIS ALREADY HAPPENED. I CAN'T SAVE YOU. NOBODY CAN."

We're also all responsible because we've created pathways for pandemics through our consumption habits. As David Quammen, who has written on pandemics, says, "What we eat, what we wear, all the other kinds of things that we buy, whether we have a cellphone or not, how many children we have (if we have children), how much we travel—all of those choices put varying degrees of pressure on the rest of the natural world."[10]

We've trapped ourselves by creating a world where it's difficult to make everyday choices that don't in some way impact on the natural world and so expose ourselves to new viruses, not to mention many other environmental threats. We've been experimenting on the natural world; the consequences are now escaping from their cages. It isn't just 'science', it's our model of human civilization. As Cole says bleakly at one point, "Look at them. They're just asking for it. Maybe the human race deserves to be wiped out." To Peter Daszak of the EcoHealth Alliance, a nonprofit that aims to protect the public from the emergence of diseases (and who has studied how infectious disease events have risen significantly), "Nature didn't send us this message. We sent it to ourselves."[11]

As the poster for *12 Monkeys* put it, 'The future is history.' Meaning, presumably, both that the future is trashed, but also that what will happen has been determined in the past. It's true in our world as well. While our actions today and tomorrow still matter, the trends that we'll describe in this book are largely already set and their effects are unavoidable. Most obviously, we tend to think of climate change. Even though mainstream media typically frames climate change as something we can still, if not avoid as such, then avert to some significant degree, actually 'all' that's left for us to decide is how highly destructive it'll be.

Echoing recurring dystopian visions, scientific studies are repeatedly setting out ever worsening scenarios. At present, it seems as if we'll probably end up with a global average temperature rise of between 2.6 and 3.9 degrees Celsius, if not more. This is at the middle to upper range of models that were set out back in a landmark climate change report in 1979 – *more than forty years ago*.[12]

In 2018, the Intergovernmental Panel on Climate Change determined we need to hold the temperature rise to 1.5 degrees Celsius to prevent catastrophic changes. As climate reporter Kendra Pierre-Louis has commented, "Based on [the IPCC's] description the difference between 1.5°C and 2°C is basically the difference between *The Hunger Games* and *Mad Max*." (We'll come back to postapocalyptic visions in chapter 7). Already, as the IPCC notes, global surface temperature has reached 1.1 degrees Celsius above 1850–1900 levels in the period 2011–2020. Indeed, global surface temperature has increased faster since 1970 than in any other fifty-year period over at least the last 2,000 years. To underline the scale of the challenge to averting the climate crisis (or at least, the very worst of its effects), all we got from the unprecedented coronavirus shutdown was a 5 percent drop in global carbon emissions, whereas according to the United Nations' projections holding global temperature rise below 1.5 degrees Celsius would require ongoing *annual* emission reductions of 7.6 percent.

At the time of writing (in its most recent *Sixth Assessment Report* in 2023), the IPCC counsels that "There is a rapidly closing window of opportunity to secure a livable and sustainable future for all." But given current trends, such warnings are probably already history. As one of the authors of a 2020 study of likely temperature rises, Dr. Kate Marvel from the NASA Goddard Institute for Space Studies, says, "There's a tendency to try to put the perfect numbers on things, to say we have 12 years to save the planet. Honestly, we have, like, negative 30 years to save the planet."[13]

"YOU'RE NOT GOING TO SAVE THE WORLD, OKAY?"

Such studies, and the IPCC reports, also typically don't take into account potentially runaway feedback loops that could create a self-reinforcing cycle of warming that could make parts of the earth uninhabitable, flood coastal cities, dramatically reduce crop yields, and result in suffering and death for billions of people (perhaps the survivors will have to relocate underground). Further, even if humanity stopped all carbon dioxide emissions today, warming would continue for hundreds of years, and several of its effects are irreversible.

It's not just climate change, of course. Ecosystem collapse is already happening as well and is similarly largely irreversible. Humans are exploiting and destroying nature on unprecedented scale. Wildlife populations are in freefall, driven by overconsumption, population growth, and intensive agriculture. According to the WWF and Zoological Society of London's biennial *Living Planet Report 2020*, global populations of mammals, birds, fish, amphibians, and reptiles fell by 68 percent between 1970 and 2016. Just two years earlier, the figure stood at 60 percent. In other words, collapse is accelerating.

The epidemic is industrial capitalist civilization—the insatiable drive for profit and interminable expansion of value which demands ever-increasing levels of extraction, production, and consumption, and which is causing massive deforestation, the transformation of ecologies into mono-crop plantations, the despoliation of nature in the quest for raw materials, the destruction of wildlife, rampant industrialization, and the exploitation and elimination of many indigenous peoples and their ecologically balanced worlds. Perhaps, in this sense at least, Peters is right: the epidemic is us.

At the airport, Railly spots Peters and recognizes him as Leland Goines' assistant—and as the 'apocalypse nut' who once challenged her about humanity's inherent destructiveness. Peters is at security. Checking his suitcase, a guard asks him about the vials of liquid he's carrying. Peters opens one of them and waves it under the guard's nose. This is the start of the contagion. Peters has a round-the-world ticket to San Francisco, New Orleans, Rio de Janeiro, Rome, Kinshasa, Karachi, Bangkok, and Beijing, the cities where it's known from later news reports that the outbreaks first happened—the interconnected modern world making the spread of disease easier and more rapid than ever before in our history.

Earlier, we've seen Peters handling a tray of vials filled with amber liquid. A passage from the Book of Revelations is quoted twice in the movie, referencing seven vials filled with God's wrath. Clearly, Peters is deeply nihilistic and anti-human, not a Cassandra figure but a fascistic fatalist (we'll return to the growth of conservative evangelical, even ecological, 'endism' in chapter 7). He had the right analysis, but the wrong reaction.

Cole and Railly realize they might be able to stop the virus from being released after all. Cole breaks through the security line and chases Peters through the airport, brandishing a gun. Security guards shoot Cole in the back. He falls down. Railly rushes to him, while Peters escapes to board his plane.

Close-up on a boy with blue-grey eyes. He watches as a blond woman weeps by a man's side as he lies on the ground. The boy seems particularly affected by the distress and tenderness of the woman as the man dies in her arms.

Of course, the boy is the young James Cole. Like him, we've witnessed our own death, we just didn't understand what it was. It awaits him in his future, and us too. That's what recurring depictions of dystopia really are—nightmare-memories that have haunted us for a long time. Contrary to techno-optimists, the dark future is not a self-fulfilling prophecy we could avoid if only we were more 'positive'; it's a given future reality we've already all foreseen.

"I WANT TO DO MY PART, GET US BACK ON TOP, IN CHARGE OF THE PLANET"

A sweaty but relieved Peters, safely aboard the plane with his suitcase, takes his seat next to a middle-aged woman. We've seen her before, in the future—she was one of the scientists instructing Cole on his mission. Cole stated earlier that once the virus was located, a scientist would be sent back to study it in its original unmutated form so that a vaccine could be developed in the future. But not, of course, to change the past. (Again, it turns out that Cole wasn't delusional—the authorities from his time really were tracking him.)

As if just speaking to herself, the woman remarks, "It's obscene, all the violence, all the lunacy. Shootings even at airports now. You might say that we're the next endangered species . . . human beings." An unsuspecting Peters agrees, "I think you're right, ma'am. I think you've hit the nail on the head." She introduces herself: "Jones is my name. I'm in insurance."

It's one of the most cryptic endings in film history. 'Insurance' could mean a backup, a second chance. The plague has begun, but the scientists still have an opportunity to seize a sample of the virus in order to produce a vaccine in the future. History can't be changed, and this wasn't Cole's mission. But the future—some kind of future—still might be.

Or maybe 'insurance' means compensation when something predictable happens. Since the scientists are from the future, perhaps they already knew Cole would fail, the pardon they promised him is meaningless, and the virus was always going to be released. It's no compensation to the dead, rather a mission on behalf of the one percent still alive in the future.

We don't know if it was ever possible to stop the release of the virus. Perhaps that would have created a time travel paradox, and since it's a paradox it would have failed. But it doesn't seem like they tried, even though, as we can see, it's also possible for them to travel into the past, seemingly easier than it is for Cole (sharp-eyed viewers might notice other scientists from the future popping up in some earlier scenes set in the past as well). And they do seem to have known, eventually, who actually did release the virus. And so perhaps the scientists rather *like* the dystopia over which they rule, aided by emergency measures, a world in lockdown. Remember, it's the future scientist who remarks that humans are an endangered species, seemingly agreeing with Peters' nihilistic world view.

Or maybe, most horrifyingly, 'insurance' means that she's there to make sure the apocalypse happens the way it's *supposed* to: Cole and Railly effectively follow a false trail focusing on the Army of the Twelve Monkeys, they fail to stop the release of the virus, Cole dies trying, and the scientists can (continue to) rule a decimated, plagued future. After all, by being sent into the past, Cole inadvertently sets the whole catastrophe in motion: he's the first to introduce the story of a virus that wipes out humanity, and he inspires Railly's initial dismissal of doomsday prophets, which in turn motivates Peters to respond with his views about the environmental destructiveness of humankind. Cole even provides the dramatic, tragic distraction at the airport that allows Peters to escape.

So, maybe the end of the world was effectively man-made. Maybe they knew what would happen all along. Maybe they didn't try to stop it. Maybe, rather than worrying about warnings of impending doom, they even welcomed it.

NOTES

1. This principle was reaffirmed in a later scientific paper, co-authored by Novikov, which is so incomprehensible to us laypeople that it almost reads like a parody (one of the co-authors of the paper, Kip Thorn, was the scientific advisor on Christopher Nolan's films *Interstellar* and *Tenet*, both of which seem to accord to the same principle): John Friedman, Michael S. Morris, Igor D. Novikov, Fernando Echeverria, Gunnar Klinkhammer, Kip S. Thorne, and Ulvi Yurtsever, "Cauchy Problem in Spacetimes with Closed Timelike Curves," *Physical Review D* 42, no. 6 (1990): 1915–1930. https://authors.library.caltech.edu/3737/.

2. Abraham Riesman, "12 Monkeys Is the Apocalypse Movie We Need Right Now," *Vulture*, March 13, 2020. www.vulture.com/2020/03/12-monkeys-why-terry-gilliams-movie-is-so-relevant-today.html.

3. Susanne Kord, *12 Monkeys* (Oxford/Liverpool: Auteur Publishing in partnership with Liverpool University Press, 2019).

4. Susanne Kord, "Pandemic Movies: What Terry Gilliam's '12 Monkeys' Tells Us About the Future," *Liverpool University Press Blog*, June 17, 2020. https://liverpooluniversitypress.blog/2020/06/17/pandemic-movies-what-terry-gilliams-12-monkeys-tells-us-about-the-future/.

5. Ibid.

6. Abdul El-Sayed, "Coronavirus Is Exploiting an Underlying Condition: Our Epidemic of Insecurity," *The Guardian*, March 19, 2020.

7. Mike Davis, "The Coronavirus Crisis Is a Monster Fueled by Capitalism," *In These Times*, March 20, 2020. http://inthesetimes.com/article/22394/coronavirus-crisis-capitalism-covid-19-monster-mike-davis.

8. Ibid.

9. Colin J. Carlson, Gregory F. Albery, Cory Merow, Christopher H. Trisos, Casey M. Zipfel, Evan A. Eskew, Kevin J. Olival, Noam Ross, and Shweta Bansal, "Climate Change will Drive Novel Cross-Species Viral Transmission," *bioRXiv* (2020). www.biorxiv.org/content/10.1101/2020.01.24.918755v1.

10. Quoted in Jonah M. Kessel, "How to Stop the Next Pandemic," *The New York Times*, September 1, 2020. See also David Quammen, *Spillover: Animal Infections and the Next Human Pandemic* (New York: W. W. Norton & Company, 2012).

11. Quoted in Jonah M. Kessel, "How to Stop the Next Pandemic," *The New York Times*, September 1, 2020.

12. National Research Council, *Carbon Dioxide and Climate: A Scientific Assessment* (Washington, D.C.: The National Academies Press, 1979).

13. S. C. Sherwood, M. J. Webb, J. D. Annan, K. C. Armour, P. M. Forster, J. C. Hargreaves, G. Hegerl, S. A. Klein, K. D. Marvel, E. J. Rohling, M. Watanabe, T. Andrews, P. Braconnot, C. S. Bretherton, G. L. Foster, Z. Hausfather, A. S. von der Heydt, R. Knutti, T. Mauritsen, J. R. Norris, C. Proistosescu, M. Rugenstein, G. A. Schmidt, K. B. Tokarska, and M. D. Zelinka, "An Assessment of Earth's Climate Sensitivity Using Multiple Lines of Evidence," *Reviews of Geophysics* 58, no. 4, December, 2020. https://agupubs.onlinelibrary.wiley.com/doi/full/10.1029/2019RG000678.

PART II

"There Was a World Once, You Punk"

Chapter 4

"They're Running Out of the Damn Green Again!"

Why Soylent Green *and the Serious Sci-Fi of the 1970s Predicted Social Collapse*

Soylent Green (1973) is the motherlode of 1970s moviemaking, encompassing dystopian sci-fi, environmentalism, urban malaise, social unrest, political corruption, and dark conspiracy. The 1970s saw a series of socially critical, and sometimes dystopian, science fiction films. *Soylent Green*'s setting reflected a growing environmental consciousness over the previous decade, from Rachel Carson's *Silent Spring* to a raft of new environmental regulations and the first Earth Day in 1970. Crucially, these concerns were shared by conservatives. But then a divide began to develop, between the new left's adoption of a critical environmental conscious that pointed to the destructiveness of industrial capitalism, and the emerging new right that identified the threat from this consciousness and fought back fiercely. Like the shocking truth revealed in the famous ending of *Soylent Green*, the truth of industrial civilization was too uncomfortable to face.

"HOW DID WE COME TO THIS?"

It begins with a montage of archive photographs, charming scenes of people enjoying nature, accompanied by a slow waltz. Then there's industrialization, a sense of optimism and progress, the music speeding up. The population grows rapidly, smokestacks, cars rolling off production lines, urban crowds, affluence, and consumption. Followed by pollution, overcrowding, war, famine, disease, the screen splitting into multiple images at once, the music now frenetic. Finally, the music slows down, over wastelands, destroyed forests, barren industrial sites, ending on a thickly polluted New York cityscape. It's the story of the twentieth century and into the twenty-first century—all in two minutes.

The opening titles by Charles Braverman are among the best ever made in establishing the context for the film—a dirty, broken future—and how we got here. *Soylent Green* is an early 1970s ecological dystopian detective conspiracy thriller (there's a combination), directed by Richard Fleischer and starring Charlton Heston, Leigh Taylor-Young, and Edward G. Robinson. It's loosely based on the 1966 novel *Make Room! Make Room!* by Harry Harrison, who wasn't at all happy with the substantial changes made to his story by screenwriter Stanley R. Greenberg. But what was kept was a dystopian future of dying oceans, year-round humidity due to the greenhouse effect, overpopulation, pollution, poverty, and depleted resources (Richard H. Kline's photography uses hazy green filters to vividly evoke a poisoned atmosphere). Welcome to 2022.

Soylent Green wasn't the only one. The 1970s saw a series of socially critical, and sometimes dystopian, science fiction films, many of which seemed to share a sense of urgent alarm about the future we were heading towards. What was going on?

To the author, architect, and academic Douglas Murphy (*Last Futures: Nature, Technology, and the End of Architecture*), what happened in the 1960s and 1970s was possibly the last chance we had of creating a decent and environmentally sustainable society. The radical movements that developed in these decades—the writer Kirkpatrick Sale called them the "awakened generation" – believed that other futures and new ways of living were possible, even inevitable.

But Murphy also emphasizes that this was a generation that had a strong sense of disaster. In the late 1960s the world was faced with impending doom: it was the height of the Cold War, the coming end of oil (it was thought), and the seemingly irreversible decline of great cities around the world. Out of these crises came a new generation that hoped to build a better future, influenced by visions of geodesic domes, walkable cities, and a meaningful connection with nature. Scenarios of doom and gloom were accompanied by radical but serious ideas to solve the problems we faced (most of which would be largely unrealized futures, it would turn out). This is the story of this chapter—of this era's most interesting sci-fi movies, the analyses that inspired them, and the warnings that we needed to change course before it was too late.

"HOW COULD I . . . HOW COULD I EVER IMAGINE . . . ?"

There had been a handful of eco-disaster films before *Soylent Green*, mainly monster movies, but none were as relevant or urgent as Fleischer's film, and none presented so bleak a picture of the human misery that could result from overpopulation, global warming, and food and energy shortages (this is a film made in the early 1970s that includes climate change, so we really, *really* can't say we weren't warned).

Soylent Green received mixed reactions from critics, however. Alan R. Howard in *The Hollywood Reporter* called it a "terrifying vision of the future made all the more urgent by today's inflationary food prices and fast approaching energy crisis." Roger Ebert gave the film three stars out of four, calling it "a good, solid science-fiction movie, and a little more." But he did also say that "*Soylent Green*'s real achievement is to create a 21st Century world that's convincing as reality; we somehow don't feel we're in a [sci-fi] picture. What director Fleischer and his technicians have done is to assume a very basic (and depressing) probability: that by the year 2022, New York will look essentially as it does now, only 49 years older and more run-down."

Penelope Gilliatt of *The New Yorker* (the other film critic at the magazine) was as negative as Pauline Kael probably would have been, writing, "This pompously prophetic thing of a film hasn't a brain in its beanbag. Where is democracy? Where is the popular vote? Where is women's lib? Where are the uprising poor, who would have suspected what was happening in a moment?" (well, we'll come back to these questions in chapter 7).

Whatever they thought of its dramatic properties, most reviewers did at least recognize that the film reflected then-current trends—as we'll see, trends that some people preferred weren't acknowledged at all.

Indeed, *Soylent Green* and many other 1970s sci-fi movies might not have been made except for two commercially (and eventually, critically) successful films that came just a few years before, in 1968. In different ways, they demonstrated how science fiction on film could address civilizational issues. Viewed from the perspective of the movies that followed them, they reflected an increasing questioning of humanity's progress and its slim chances of survival. These films were the pivot between the 1950s style sci-fi that counselled responsibility with regards to our rapidly expanding scientific powers, and the grounded, almost-here dystopian sci-fi of the 1970s that assumed we'd make foolish choices.

Planet of the Apes was hugely popular and spawned four sequels, a live action and animated TV series (and more recently, a pretty good trilogy of new films) – the first successful sci-fi franchise. Adapting French author

Pierre Boulle's 1963 novel *La Planète des Singes* (*Monkey Planet* in the UK), which was heavily influenced by eighteenth and nineteenth-century utopian-dystopian travel narratives especially Jonathan Swift's satirical *Gulliver's Travels*, American astronauts discover a future world ruled by advanced apes, who own humans as slaves and use them in scientific experiments.

Entries in the series became more explicitly political as they went on, especially regarding racial issues and conflict. Bizarrely, the first film's producers were unaware of its racial themes until Sammy Davis Jr. pointed them out. But as Eric Greene (*Planet of the Apes as American Myth: Race, Politics, and Popular Culture*) suggests, sci-fi's distance can provide deniability for filmmakers and audiences, making issues less psychologically and politically challenging on first viewing. The *Planet of the Apes* films are also about humanity's arrogant but fragile superiority over animals and the environment, and our likely humbling (as suggested by the first film's famous final scene) if we don't change our ways.

Apes pop up again at the start of *2001: A Space Odyssey.* At least this film suggests we'll be saved by intelligence, but unfortunately it's not our own. After the discovery of an alien monolith on the moon, astronauts voyage to Jupiter to locate its source. Written by director Stanley Kubrick and author Arthur C. Clarke, the story encompasses evolution, existentialism, technology, artificial intelligence, and extraterrestrial life.

The spaceship's seemingly coolly composed computer, H.A.L., has a breakdown, the result of 'his' struggle with the dehumanized crew's duplicity. Earlier versions of the script were clearer that Cold War tensions had led to nuclear weapon satellites circling Earth (Kubrick discarded them because of the thematic similarity to his previous film *Dr. Strangelove*). *2001* only really took off at the box office when younger people seized on its trippy stargate ending in which the sole surviving astronaut encounters . . . something wonderful.

Once again, let's see what Pauline Kael thought. Like many critics at the time, she dismissed *2001* as a "monumentally unimaginative movie" and derided its popularity with young people as part of a "new tribalism", a counterculture fad. To Kael, the film was profoundly anti-human, it "celebrates the end of man." At least she thought *Planet of the Apes* was "one of the best science-fiction fantasies ever to come out of Hollywood." That wasn't great praise, though, since Kael didn't think sci-fi produced much "literary art." And she disliked its point-scoring against "establishment thinking": "These swifties are not Swift, and the movie's posture of moral superiority is somewhat embarrassing."

But Kael did recognize the smart casting of Charlton Heston as the hero: "He represents American power . . . the perfect American Adam to work off some American guilt feelings or self-hatred on, and this is part of what makes this new violent fantasy so successful as a comedy." We'll come back to Heston in a moment.

"SOMETHING STINKS HERE"

Both films were prescient, but they also came at the right time, before the onset of disenchantment. After the 1969 moon landing, public interest in the space program began to wane (Kubrick had raced to get his film out before the Eagle had landed). The next wave of sci-fi movies would address issues much closer to home. More serious, thought-provoking science fiction films were released in the 1970s than in the forty years before, and arguably ever since. *Soylent Green* was just one of a slew of American sci-fi films in the 1970s that, more than imagining the future, reflected the present and its problems, which is of course what sci-fi can be so good at doing. Among them were *Colossus: The Forbin Project* (1970), *The Omega Man* (1971), *THX-1138* (1971), *The Andromeda Strain* (1971), *Silent Running* (1972), *Westworld* (1973), *Zardoz* (1973), *Rollerball* (1975), *A Boy and His Dog* (1975), *The Stepford Wives* (1975), *Logan's Run* (1976), *The Man Who Fell to Earth* (1976), *Invasion of the Body Snatchers* (1978), *Quintet* (1979), and *Alien* (1979) (we'll discuss *Colossus* and *Rollerball* in later chapters).

The so-called new Hollywood took advantage of sci-fi, among other genres, to tell different stories, address social issues of the time, and explore new perspectives. Younger American filmmakers also responded to storytelling and stylistic developments in European cinema, notably the French New Wave, treating their films as cinematic statements and social commentary, reflecting their artistic vision and broader worldview.

Not everyone enjoyed this output, however. Only a few of these films were commercially successful. And some commentators bemoaned the inward turn in science fiction film to America's increasing social and political problems. For example, writing in 1978, Joan F. Dean complained "all that the science fiction films of the early seventies offer in the way of [traditional sci-fi staple] extra-terrestrial life were bacteria, David Bowie, an invisible civilization, and a few perverts."[1] Dean dismissively regarded the era as witnessing a dearth of futuristic films, but we remember it now as a particularly rich period of social seriousness in mainstream sci-fi, in large part because of what was going on here at home.

"I KNOW. SOL. YOU TOLD ME BEFORE. A HEAT WAVE ALL YEAR LONG. A GREENHOUSE EFFECT. EVERYTHING IS BURNING UP"

In *The Omega Man*, another Charlton Heston dystopia movie from a couple of years before, the civilization-ending plague turned out to be human beings (at least, this was clearer in Richard Matheson's 1954 novel *I Am Legend*, on which the film was based).

Similarly, in *Soylent Green*, we've also screwed up the world, and our means of survival. In its New York City of 2022, the bounty of the world, which once seemed inexhaustible, has dried up. The temperature never drops below 90 degrees Fahrenheit. Combined with population growth and pollution, climate change has all but eliminated the supply of natural foodstuffs and resources (although the actual term 'global warming' would only be coined two years later by scientist Wallace Broecker).

The majority of the population rely on processed 'food' rations produced by Soylent Industries, a conglomerate that controls two-thirds of the world's food supply and manufactures cheap crackers out of chemicals and vegetable matter. People queue in long lines for Soylent Red, Soylent Yellow, and especially the new wonder food stuff that will save humanity, the protein-rich Soylent Green. These are small wafers advertised as made from "high-energy plankton", which are much more nutritious and palatable than Red and Yellow. However, like other foods, Green is in short supply, which leads to regular riots. Protestors are brutally removed from the streets by police vehicles that scoop them up with large shovels and dump them inside the vehicle's container.

Corporations dominating society was a 1970s story trope, but it also anticipates future sci-fi genres such as cyberpunk. And as John Farrell puts it: "This is the price of a mollified, ignorant populace who depends on their subsistence from a monolithic entity that views them as nothing more than commodities and customers. This, whether it looks like it or not, is cyberpunk."[2]

Soylent Green's setting also reflected a growing environmental consciousness over the previous decade. Rachel Carson's *Silent Spring*, published in 1962, is remembered now as an indictment of the chemical pesticide DDT, but it's really a critique of postwar faith in scientific progress, the idea of human dominance over the environment, our dependence on industrial food production, and an assumed culture of abundance.

Carson's impact was in part due to her ability to engage non-scientific audiences. *Silent Spring* actually began with a piece of speculative fiction, 'A Fable for Tomorrow', which imagined a once bountiful town hit by a "strange blight" that causes animals to die and people to catch diseases—a dystopian "stricken world" of endless disasters. In response, the Monsanto Corporation vainly tried to defend its business with its own work of fiction 'The Desolate Year' (asking 'what if pesticides didn't exist?'), but no-one remembers that now.

In 1967, sparked by Carson's work, the Environmental Defense Fund began bringing lawsuits against US states for citizens' rights to a clean environment. Then came a raft of new regulations and institutions—in 1969 the National Environmental Policy Act which introduced a new requirement for environmental impact statements, and in 1970 the creation of the Environmental Protection Agency (EPA).

These weren't the only concerns. The emerging environmental movement was also spurred by increasing exploration and drilling in environmentally sensitive areas. Air pollution was another important factor; in many ways, it was central to the growing environmental movement. In 1963, Congress had enacted the first Clean Air Act, an acknowledgement that smog had become a national problem. Two years later, it called for the first national emissions standards for cars. In 1969, the Justice Department sued automakers for conspiring to delay anti-smog devices (ultimately settled out of court). Then Congress enacted the law that has set the framework for air pollution regulation, the Clean Air Act of 1970, supported by both Democrats and Republicans. There was also the Clean Water Act and the Endangered Species Act—all under a Republican president, Richard Nixon.

The focus on air quality also spurred an increasing understanding of carbon dioxide. The very first prediction that the planet would warm as humans released more of the gas was made back in 1896. In 1965, President Lyndon Johnson's Science Advisory Committee on Environmental Pollution noted the greenhouse effect caused by carbon dioxide, and blamed the unwitting "vast geophysical experiment" being conducted as a result of "worldwide industrial civilization." It predicted warming oceans, rising sea levels, and other disasters if nothing was done. It's only subsequent propaganda that makes us believe that 'we didn't really know back then.'

As has been revealed, the fossil fuel industry knew full well—more than almost anyone else—what the consequences of this 'experiment' could be. For example, in 1977 Exxon scientists and managers were warned that doubling carbon dioxide concentration in the atmosphere would increase average global temperatures by 2 to 3 degrees Celsius, and as much as 10 degrees Celsius at the poles.[3] Rainfall would get heavier in some regions, while other places might turn to desert; some countries would see their agricultural output reduced or destroyed. Of course, Exxon and other oil companies would deny, obfuscate, and delay action on climate change, and their campaign would start earlier than has often been thought—from the late 1970s/early 1980s, rather than the late 1980s.

"YOU KNOW, THERE ARE 20 MILLION GUYS OUT OF WORK IN MANHATTAN ALONE JUST WAITING FOR MY JOB AND YOURS TOO"

After fighting an ape-dominated future and surviving a largely people-less one, in *Soylent Green* Charlton Heston struggles through a heavily overcrowded tomorrow. He plays NYPD detective Frank Thorn who, given population pressures, lives with his aged friend Solomon 'Sol' Roth (Edward G. Robinson), a highly intelligent analyst referred to as a 'Book.'

Sol is a former college professor whose job is to sort through the remnants of written records to help Thorn with his investigations. Sol repeatedly says he remembers the world before ecological disaster and the population crisis, when there were animals and real food. Thorn finds Sol's utopian-sounding stories somewhat hard to believe. Anyway, Thorn is tasked with investigating the murder of the wealthy William R. Simonson, apparently by a burglar, and quickly learns that Simonson was a board member of Soylent Industries.

Soylent Green is the motherlode of 1970s moviemaking, encompassing dystopian sci-fi, environmentalism, urban malaise, social unrest, political corruption, and dark conspiracy—not the kind of thing you'd normally associate with Charlton Heston. As Kael noted, Heston was Mr. America, later known for his staunchly conservative politics, including his prominent presidency of the National Rifle Association. However, he was also particularly concerned with overpopulation—a largely bipartisan issue at the time—and was instrumental in getting the film made despite studio doubts (a new ending helped to assuage their concerns that the movie might be used to advocate for birth control against religious objections).[4]

Soylent Green wasn't the only population-themed film at the time. *Z.P.G.* (standing for zero population growth), a British sci-fi movie from 1972 directed by Michael Campus, imagined a totalitarian world state that banned childbirth on penalty of death (tagline: 'The Time Is Tomorrow and There's No Time Left'). Violators are placed in a plastic bubble and suffocated in public. In fact, everything is strictly regulated, from food to living quarters to recreational activities. Yet again it's a polluted, overpopulated future (the film frequently references the 1970s as a turning point). Restaurants are crowded. Respirators are needed to walk outside. Even common animals are extinct. Couples are provided with 'lifelike' dolls, but one woman wants a real child instead. It doesn't end well.

As we're already seeing, overpopulation is a common theme in dystopian films. Indeed, *Z.P.G.* was inspired by the contentious 1968 nonfiction book *The Population Bomb* by Paul R. Ehrlich and (an uncredited) Anne Ehrlich. This predicted famine, overcrowding, and other social problems as the population grew unchecked. The book opened with the memorably apocalyptic lines, "The battle to feed all humanity is over. In the 1970s the world will undergo famines—hundreds of millions of people are going to starve to death in spite of any crash programs embarked upon now." And it only got worse from there.

The Ehrlichs' bestseller was published in paperback by the Sierra Club, the environmental group, in partnership with Ballantine Books, known for its politically engaged science fiction alongside serious nonfiction dealing with social issues such as contraception, nuclear energy, and the environment. The Ehrlichs presented readers with three sci-fi style "doomsday scenarios" combining mass famine, plague, and nuclear war; they later claimed that the "biggest tactical error in [the book] was the use of scenarios, stories designed to help one think about the future." But this was also what helped to ensure the book's phenomenal success, as well as its enduring controversy.

Indeed, the use of science fiction was part of a strategy widely adopted not only by individual authors but also organizations such as Zero Population Growth (ZPG), cofounded by Ehrlich in the same year *The Population Bomb* was published. For example, ZPG edited and supported a 1971 anthology of sci-fi short stories devoted to overpopulation by writers such as J. G. Ballard, Ray Bradbury, and Doris Lessing called *Voyages: Scenarios for a Ship Called Earth* (we'll return to that metaphor in a moment).

So, there was a thriving subgenre of science fiction, dubbed demographic-dystopian or 'demodystopian.' Paul Ehrlich had endorsed Harry Harrison's *Make Room! Make Room!* as "the most effective fictional treatment of the consequences of the population explosion that I have ever come across." But *Z.P.G.* the film fared poorly at the box office and with critics. It also became embroiled in a major row with Ehrlich's Zero Population Growth over its title, and because of its dramatized method of mandatory population control.

Nonetheless, population growth was a mainstream political concern at the time. As president, Lyndon Johnson expressed concern about the "explosion" in population and "growing scarcity" of world resources. From 1965 to 1967, he oversaw a revolution in federal policy towards birth control and population planning. Later, Richard Nixon created a Commission on Population Growth and the American Future. But as we'll see in the next chapter, as the 1970s wore on the desire for economic growth would trump population concerns. And as we'll also see, the apocalyptic over-claiming in the Ehrlichs' work would prove helpful to the conservative counter-revolution to come; as in *12 Monkeys*, catastrophe-warning Cassandras can often be easily dismissed.

"PEOPLE WERE ALWAYS ROTTEN. BUT THE WORLD WAS BEAUTIFUL."

But the real problem was that works such as *The Population Bomb* distracted from the broader, deeper problems that industrial capitalist civilization was creating—problems that *Soylent Green*, with its sum-of-all crises backstory, *did* point to.

In *Soylent Green* it's not just population that's the problem. In its New York, the masses are mainly unemployed, dirt poor, and live in abandoned cars, tent cities, and the hallways of buildings. Only the city's elite can afford spacious, air-conditioned apartments, clean water, and natural food, the latter at horrendously high black-market prices. Their homes also include concubines referred to as "furniture" – domestic (sex) slaves.

At the crime scene in Simonson's luxury high-rise apartment, Thorn talks with the building superintendent about discovering the body. Instead of looking for clues, the poorly paid detective helps himself to the wealthy man's food, liquor, and books, and even enjoys a hot shower! Only one-percenters like Simonson can afford real strawberry jam, beef, and apples. When Thorn pilfers them from his apartment, Sol is overcome with emotion, remembering a time when such things were taken for granted.

Thorn also questions Shirl (Leigh Taylor-Young), the apartment's attractive young concubine, and Simonson's bodyguard, Fielding, who says he was escorting her on a shopping trip when the murder happened. Returning home, Thorn gives Sol a copy of the *Soylent Oceanographic Survey Report, 2015 to 2019* taken from Simonson's apartment.

Other sci-fi from the 1970s moved the environment from the background to front and center. A year before *Soylent Green*, *Silent Running* encapsulated the idea of 'spaceship Earth', which had become a key environmental concept. Unlike some 1970s avant-garde influenced (or merely entertainment-first) sci-fi, it was absolutely clear, some would say browbeatingly so, what *Silent Running* was saying.

Directed by Douglas Trumbull, who created the groundbreaking effects for *2001*, *Close Encounters of the Third Kind*, and *Blade Runner* among others, *Silent Running* is about the last forests of a barren (and again, overpopulated) Earth preserved in large geodesic domes aboard spaceships orbiting Saturn. When the order is given to destroy the forests and return the ships to commercial service, one of the astronauts (Bruce Dern), kills his crewmates and hijacks the spaceship, making off with the last surviving biodome.

Trumbull's movie was inspired by the idea that smog and light pollution in Los Angeles had caused people to become disconnected from the stars they could no longer see. To Trumbull, how we care for the Earth's environment is connected to how we view the universe, so an ecological movie set in space was a natural way to connect these issues.

In 1968, Apollo 8 had been the first crewed spacecraft to leave Earth orbit, reach the Moon, and return. The astronauts orbited the Moon ten times, during which they made a Christmas Eve broadcast, including reading from the Book of Genesis. At the time it was the most watched TV program ever. The mission also produced the first image, probably taken by William Anders, of the whole Earth floating in space—the famous 'Earthrise' photograph which prompted a new understanding of the planet as one fragile ecosystem.

In February 1970, *Time* magazine ran a lengthy cover story titled 'Fighting to Save The Earth from Man.' The year before there had been a massive oil spill in Santa Barbara, California, motivating Wisconsin Senator Gaylord Nelson to come up with the idea for the first Earth Day, which took place in April 1970. Earthrise was another inspiration. Twenty million people participated in Earth Day events—the so-called Earth Day generation.[5]

"YOU KNOW WHAT I REALLY THINK, LIEUTENANT? I THINK IT'S REALLY BUSTED THIS TIME."

Silent Running's biodomes represent miniature spaceship Earths: closed spaces for nature (and humans) to survive with a limited carrying capacity, a tight 'economy' requiring careful and respectful management. The metaphor of spaceship Earth had actually been around for a few years before this, for example the book by political scientist and economist Barbara Ward, *Spaceship Earth* (1966). In 1969, the architect, inventor, systems theorist, and futurist Buckminster Fuller wrote his essay *Operating Manual for Spaceship Earth*, arguing that humans would have to become the operators of our own life support and maintenance systems. Similarly, in 1971, Paul Ehrlich and Richard L. Harriman published *How to Be a Survivor: A Plan to Save Spaceship Earth*.

But reflecting the dour (some might also say defiant) ending of *Silent Running*, dystopian pessimism was setting in, echoed in books such as *Eco-Catastrophe* (the Editors of Ramparts, 1970), *Terracide: America's Destruction of Her Living Environment* (Ron M. Linton, 1970), the very 1970s-sounding *Where Have All the Flowers, Fishes, Birds, Trees, Water, and Air Gone?* (Osborn Segerberg, 1971), and *The Last Days of Mankind: Ecological Survival or Extinction* (Samuel Mines, 1971).

Relatedly, the 1970s also saw the back-to-the-land movement. Almost one million urban Americans relocated to rural homesteads to practice forms of self-sufficiency. Many, but by no means all, were white liberals who adopted voluntary simplicity as a result of their environmental beliefs, to get closer to nature, and grow and eat wholesome food. Some of them were guided by *The Whole Earth Catalog*, a countercultural and product review magazine published by Stewart Brand between 1968 and 1972 and occasionally thereafter, which focused on self-sufficiency, ecology, alternative education, do it yourself, and holistic thinking. Some feared a forthcoming economic collapse. As Jeffrey Jacob notes, "the back-to-the-land movement was, in its own quiet way, a broad-based protest against what the spirit of the sixties saw as the irrational materialism of urban life."[6]

Spaceship Earth's ecosystem was also reflected the growth in 'new economics.' For example, in 1966 Kenneth Boulding had spoken about the "economics of the coming spaceship earth" signifying the "transition from the illimitable plane to the closed sphere." (It was Boulding who also famously said that "Anyone who believes exponential growth can go on forever in a finite world is either a madman or an economist.") Earlier generations had the conception of there always being somewhere beyond their limited world. Now, the reckless, exploitative, violent "cowboy economy" of the endless plains needed to be replaced by a carefully cultivated "spaceman economy" – a new economy of circulation, exchange, and renewal in the "closed earth of the future."

Although to begin with few in the new left (such as student protestors against the Vietnam War) were environmentalists, they shared a number of themes with this growing green thinking: criticism of a machine-like industrial-political system leading us to destruction, the madness of the establishment, and the need for a new cultural consciousness. Leaders such as Robert Kennedy and Dr. Martin Luther King criticized endless growth and consumerist materialism. But there were tensions as well; some activists and campaigners dismissed environmentalism as merely an 'anti-litter campaign', and an elitist distraction from the struggle for racial equality, and against poverty, war, and imperialism (indeed, some even regarded it as racist because of its concerns about population growth).

Leftist intellectuals such as Herbert Marcuse, Murray Bookchin, and Paul Goodman, among others, advanced critiques of industrial society and its dysfunctional relationship with an exploited nature and our alienated selves. The new left was fundamentally humanist, including a critique of industrialization, which some thinkers regarded both 'liberal' capitalism and old left statist communism as being dependent on (as well as bureaucracy, militarism, an attachment to technological solutions, and centralization). However, Bookchin's 'post-scarcity anarchism' saw technology as liberatory *if* it was used in an ecologically sensitive manner to produce a society of abundance for all; unlike some of his contemporaries, Bookchin disagreed with the notion of limited resources, which we'll discuss later. In their various ways, these thinkers looked back to a pre-Marxist utopian socialism that was decentralized, worker controlled, participatory, pluralist, and less materialistic.

Other environmental economists and thinkers, such as Herman Daly, E. F. Schumacher, and Barry Commoner, critiqued technological progress and growth and argued for smaller-scale, post-growth, more sustainable living. Schumacher's 1973 book *Small Is Beautiful*, which promoted a "maximum of well-being with a minimum of consumption," was another bestseller; its author was invited to the White House by Jimmy Carter.

"YOU'RE BOUGHT AS SOON AS THEY PAY YOU A SALARY"

One popular book that brought together the growing environmental consciousness and the counterculture was Charles Reich's *The Greening of America* (1970). Subtitled *How the Youth Revolution Is Trying to Make America Livable*, it first came to prominence as a lengthy excerpt in *The New Yorker*. It soon became a number-one nonfiction bestseller, dominating *The New York Times* list for thirty-six weeks and going on to sell two million copies.

In the fashion of the era, Reich argued that states of mind, not social or economic classes, are fundamental to politics in the United States. He described how the self-reliant individualism of early America had given way to the status driven conformity of corporations, criticized consumerism as a passive, unsatisfying form of self-expression, and explained how the corporate state fails to protect the environment. With the rigor of an intellectual (Reich was a legal academic) but the enthusiasm of a young person, *The Greening of America* criticized the dominant economic and political institutions in American life and celebrated the counterculture for its new and different ways of thinking.

What Reich called 'Consciousness I' sees society as an economic struggle between self-interested individuals competing in a world of scarcity for material wealth and power. It defines progress in solely economic terms, in the greater production of goods. In contrast, 'Consciousness II' recognizes that competition between individuals in the market has been replaced by rivalry between large organizations. The way to get ahead is to join a large organization and dedicate yourself to its goals, whatever they might be. But its definition of 'progress' is similarly solely material.

'Consciousness III' is the reaction to these. A person with Consciousness III is aware that institutions, whether private or public, have been organized into an all-powerful monolithic structure which knows only "the value of technology-organization-efficiency-growth-progress." To keep going, this 'machine' requires a society composed of people who are willing to work in order to consume what it produces. Instead, "Consciousness III, youth consciousness, is the key to our future, simultaneously self-directed and communitarian, adventurous, anti-hierarchical, and above all open to continuous change."

Like many others, and echoing what we discussed in chapter 2, Reich thought the machine was malfunctioning: "I see self-destruction now on a grand scale. That is, the unwillingness to pay for the things society needs. That's the most basic kind of self-destruction. That we're not prepared to pay for schools, we're not prepared to pay for highways. That is self-destruction. What are we doing to ourselves? It is nuts."[7]

The Greening of America captured and connected with a particular political-cultural moment, just before it began to recede. As we'll see in chapter 6, such was the book's popularity it was referenced in one influential strategy which urged conservatives and corporations to fight back against the rising environmental consciousness.

Reich received a lot of criticism as well as acclaim. It's even possible to interpret another well-known 1970s sci-fi movie as a dystopian rejoinder to his argument. *Logan's Run* (1976), directed by Michael Anderson and based on a book by William F. Nolan and George Clayton Johnson, is set in a massive dome city in a postapocalyptic 2274, following war, social unrest, and environmental devastation. But it's a utopia for twentysomethings: the young citizens live hedonistic lifestyles and a computer takes care of all of their needs. However, to prevent overpopulation, everyone must undergo the public rite of 'Carousel' when they reach the age of 30, when they're killed under the guise of being 'renewed.'

At first sight *Logan's Run* might appear rather new leftish: advanced capitalism manipulates us with consumer abundance and entertainment-oriented lifestyles, but we're ultimately disenfranchised by those in control. But it's possible to read the movie as a criticism of the youth culture that Reich celebrated and its perceived sense of entitlement: instead of saving the world, young people's 'extreme' thinking would produce a dystopian regime that (literally) cancels older people in the name of 'ecological balance.'

But of course, Reich saw a lot of hope in the new generation—and their ability to change their world. Unfortunately, it didn't turn out that way. Instead, as Reich (who died in 2019) later noted, "For the last 40 years, I have watched the un-greening of America." But we'll get to that.

"YOU DON'T UNDERSTAND. I'VE GOT PROOF. THEY NEED PROOF, I'VE SEEN IT."

Thorn starts a relationship with Shirl, and her help leads to a priest who Simonson had confessed to shortly before his death. Thorn goes to a local Catholic church, which is teeming with homeless people, and questions the priest about Simonson's confession. The priest is almost catatonic with exhaustion, but when he remembers Simonson, he tells Thorn he's unable to describe what Simonson said because it's "too horrible." The next day, the priest is murdered.

Thorn is told to end his investigation by his boss, on the order of the state governor, but he presses on. Clearly, he's starting to make the authorities nervous. An attempt is made on his life: he's attacked during a riot by the same man who killed Simonson, but the killer is crushed by one of the police riot control vehicles (we, but not Thorn, now know that the hit was ordered by the governor, who's running for reelection).

Meanwhile, Sol has discovered from Soylent Industries' reports that the oceans are dying—they can no longer produce plankton. The film was prophetic about this; plankton has declined more than 40 percent since 1950, and the rate of decline is accelerating, as is species extinction broadly.[8] Thorn doesn't think this is why Simonson was murdered, but he and Sol deduce that his murder must have been ordered by his fellow Soylent board members, aware that Simonson was struggling with the truth about something.

Soylent Green is set in 2022, which is the fiftieth anniversary of a report called *The Limits to Growth* (1972 was also the year of the first global environment conference, the United Nations' Conference on Human Environment, held in Stockholm). The facts contained in *The Limits to Growth* were just as threatening to powerful interests as Soylent's fictional data. And even though *The Limits to Growth* was much more levelheaded in tone than the Ehrlichs' *Population Bomb*, its authors came to be depicted in much the same way—as contemporary Cassandras, and treated just as harshly.

Also like *Soylent Green*, and despite its sobriety, *The Limits to Growth* was the motherlode. Peppered with computer-generated graphs and written in clear, dispassionate language by a team of Massachusetts Institute of Technology graduate students led by two young scholars, Dennis and Donella Meadows, it delivered a devastating conclusion: "If the present growth trends in world population, industrialization, pollution, food production, and resource depletion continue unchanged, the limits to growth on this planet will be reached sometime within the next one hundred years. The most probable result will be a rather sudden and uncontrollable decline in both population and industrial capacity."[9]

The study was initiated and funded by the Club of Rome, an ad hoc, apolitical group of scientists, academics, civil servants, and businesspeople. At the center of the club was Dr. Aurelio Peccei, a charismatic Italian industrialist and economist who had been jailed briefly in 1944 by Mussolini's security services for helping the anti-Fascist resistance. As Peccei put it, the Club of Rome's mission was to "rebel against the suicidal ignorance of the human condition." It sought to better understand the links between economics and the environment, or what it called "the world problematique" (more on which below).

The study was based on a computer simulation called World3, a 'systems dynamics' model based on the methods of MIT professor Jay Forrester, which tracked growth and the interactions between five factors: population, food production, industrial production, the consumption of nonrenewable resources, and pollution. By linking the world economy with the environment, it was the first integrated global model. Like the notion of spaceship Earth, systems dynamics recognized that closed systems involve feedback loops, inputs, and outputs that act on and affect each other. *The Limits to Growth* would put quantitative scenario analysis—possible futures based on hard numbers—into environmental studies.

Whatever its groundbreaking computer modelling, *The Limits to Growth*'s message was one that many people were intuitively aware of and could already see around them, in pollution, rising prices, plastic-tasting food, urban decline, and political unrest. *The Limits to Growth* appeared to confirm the sense of being on a civilizational road to dystopia.

"THEY CAN'T RISK, UH . . . CATASTROPHE, THEY SAY"

When Sol finally realizes the secret behind Simonson's murder, he decides to 'go home,' a euphemism for dying. He's taken into a cooled room at a government euthanasia clinic and given a potion. Listening to Rachmaninoff, he's surrounded by a wonderous film showing the Earth as it used to be, when animals, plants, forests, and oceans were thriving. It's a powerful scene, and emotionally exhausting, all the more so for knowing that Edward G. Robinson died twelve days after filming it.

Thorn returns to their apartment and finds a note from Sol. He races to the clinic and forces his way in. He can barely comprehend the images of unspoiled nature he sees. But Sol understands what he's seeing. He also knows that nature is gone and never coming back. During Sol's final moments, he tells Thorn the secret of Soylent Green, and begs him to follow his body to the 'processing center' so that Thorn can find the proof.

What was *The Limits to Growth*'s 'secret', its so-called problematique? As described by the Club, it was something apparent but still hidden that needed to be brought into the light—a detective mystery, as it were, encompassing poverty and insecurity, destruction of the environment, a loss of faith in institutions, urban spawl, alienation, and economic dislocation. These trends were happening in all societies to some degree, they contained technical, social, economic, and political elements, and most importantly, they were interacting with each other. But because we typically saw problems in isolation (and largely still do), we weren't connecting the clues.

The researchers plugged the best available data into their model and ran numerous scenarios based on different sets of assumptions. What came out was bad. Very bad. *The Limits to Growth* centered on a problematique, but it left a predicament, or rather a multitude of them. This happened because of the feedback between different global subsystems such as industry, health, and agriculture. For example, higher industrial output might mean more money to spend on healthcare and agriculture, but also more pollution, which would damage health and food production. There were no easy ways out. Apparently, even the study's researchers were overwhelmed by their findings.

However, few people now remember that *The Limits to Growth* was built around a series of scenarios, based on twelve possible futures stretching from 1972 to 2100. The main conclusion of the study was actually that delays in global decision-making would cause the human economy to overshoot planetary limits before growth in the human ecological footprint could be slowed down. *The Limits to Growth* was really a warning—about delay, denial, and about giving up and committing ourselves to a collective death.

Indeed, delay was integrated into some of the scenarios, showing how costly it would be. Of the twelve scenarios in the study, seven ended in collapse, one in half-collapse, and the rest in equilibrium. But the most realistic scenarios had the worst outcomes. Aided and abetted by a drama-hungry media, it was the collapse scenarios that were seized on by critics who dismissed *The Limits to Growth* as just another in a long line of hysterical doomsday prophecies. As we'll see in the next chapter, some critics argued fiercely that it failed to understand that human ingenuity and adaptability could prevent collapse.

Equally, there were critiques from the environmental left, for example Barry Commoner emphasized that it's *capitalism*—implied but not explicitly named in *The Limits to Growth*—that's chiefly responsible for catastrophic resource depletion, and that environmental degradation is a class issue since it disproportionately affects poorer countries and communities. 'Growth' is really the right of *elites* to consume without limits.

What critics ignored were *The Limits to Growth*'s more optimistic scenarios, such as: "It is possible to alter these growth trends and to establish a condition of ecological and economic stability that is sustainable far into the future. The state of global equilibrium could be designed so that the basic material needs of each person on earth are satisfied and each person has an equal opportunity to realize his individual human potential."

However, this kind of adaptability wasn't what critics of the study had in mind.

"SO REMEMBER – TUESDAY IS SOYLENT GREEN DAY"

Thorn sneaks into the basement of the facility, where he sees corpses being loaded onto waste disposal trucks. He secretly hitches a ride on one, which is driven to a heavily guarded processing plant just outside the city. Once inside, Thorn sees what really happens to the corpses. Thorn is spotted by workers and chased. Thorn kills one and subdues another before he escapes from the plant just as the alarm is sounded.

He heads for the Supreme Exchange, a library and gathering place for Books like Sol, to report what he's found, but he's ambushed by Fielding, Simonson's bodyguard, and three other gunmen. During a chase through the deserted streets (empty of people because of a state-imposed curfew), Thorn kills the gunmen chasing him except for Fielding, who shoots him in the back. Eventually, Thorn manages to kill Fielding as well.

When the police arrive, the wounded Thorn confides to his boss the secret of Soylent Green. As he's being carried out on a stretcher, he urges his boss to spread the word and shouts to the crowd: "Soylent Green is people! We've got to stop them . . . somehow!" Although Thorn's boss promises to make it known to the public, we suspect the truth will remain a closely guarded secret that powerful people will carry on killing to protect.

Like *Soylent Green*, *The Limits to Growth* was a hit. The study sold twelve million copies, was translated into thirty-seven languages, and remains the best-selling environmental book of all time. It made headlines in national newspapers for months, and would provoke intense debate (and much derision) for years to come. President Carter and Canadian Prime Minister (Pierre) Trudeau launched similar studies; many books would later echo its conclusions. Carter even invoked it in his inauguration speech in 1977: "We have learned that 'more' is not necessarily 'better,' that even our great Nation has its recognized limits, and that we can neither answer all questions nor solve all problems." As we'll see, the next administration would ignore such concerns. Nonetheless, after *The Limits to Growth*, environmentalists, scientists, some policymakers, and parts of the public would understand ecological problems in planetary terms and as deeply interconnected.

Much like *Soylent Green*'s famous revelation (which is pretty much revealed in its trailer), many people might think the conclusions of *The Limits to Growth* were obvious, albeit profoundly uncomfortable: collapse is unavoidable if we carry on as we were. The study's analysis hinged on the idea of infinite, exponential growth—the more a thing grows, the faster its rate of growth in absolute terms will be. Moreover, by the time humanity feels the full impact of the environmental crisis, the problem could be accelerating so fast that it might be impossible to stop, hence the study's use of scenarios to spell this out.

Soylent Green's commerce-cannibalism ending is meant to shock us into action against environmental degradation and resource depletion. It also condemns an industrial capitalism which inevitably begins to consume itself, or at least us. The economist Julian Simon, a very public critic of the Ehrlichs' population predictions, was famous for calling people "the ultimate resource", but presumably he didn't mean it in this way. The world of *Soylent Green* is though, by 'necessity', a kind of spaceship Earth closed economy: 'redundant' people are turned into food into order to feed other redundant people. Most of the scenarios in *The Limits to Growth* predicted that industrial output would start to decline in the 2020s, with population declining in the 2030s. Soylent Industries' 'solution' to the limits to growth might have helped to stave off total collapse in its fictional world—at least for a time, but maybe not for long.

Of course, *Soylent Green* is an extreme dystopia—a deliberate extrapolation. For *The Limits to Growth* however, it was the critics, not the study's authors, who would effectively make the worst outcomes inevitable. As the study suggested, the necessary equilibrium could happen only in two ways: either through a managed decline organized by humanity, or through a chaotic collapse induced by nature. The only thing that *couldn't* happen was for society to remain forever on an unsustainable trajectory, using more of nature every year than nature produces.

So in spite of its moderation, and perhaps more than its authors realized, *The Limits to Growth* had challenged something fundamental: the belief that more growth and technology inevitably means more progress. The carefully guarded secret—that more growth and technology could instead be our doom—was out. Even so, powerful people would still kill to protect it.

"YOU KNOW WHAT I REALLY THINK, LIEUTENANT? I THINK IT'S REALLY BUSTED THIS TIME"

The period of serious, mainstream, and (sometimes) popular sci-fi movies that had started with *Planet of the Apes* and *2001* and continued through the

1970s was not to last. The poor box office performance of 1980s films like *Blade Runner*, *Tron*, *The Thing*, and *Dune*, among others, marked the end of a (short) period rich with idea-driven science fiction.

There have been great sci-fi films made since, some of which we'll discuss later in this book, many of them drawing on similar themes and ideas to the ones in this era. But there hasn't really been anything like the 1970s. Part of the reason was the end of the studio era and a brief period in which younger, more experimental filmmakers took the opportunity to bring new, contemporary sensibilities to mainstream cinema.

But it was also about the political context of the 1970s, a time in which the future was still, just about, at stake—if we addressed environmental and social issues honestly and seriously, and if we really wanted to save the future. Until much more recently, in the same way films such as *Soylent Green* and others tried to do in fiction, there wouldn't be another moment in which the scale and severity of the civilizational crises we face was discussed so openly. Looking back, it's remarkable both that such films were made, and that studies such as *The Limits to Growth* gained such widespread coverage, even if much of the resulting commentary was highly critical and dismissive.

Just like later films, other books would repeat the warnings from *The Limits to Growth*, often with a stronger argument that we (still) weren't listening, for example, William R. Catton's *Overshoot: The Ecological Basis of Revolutionary Change* (1980). Catton's analysis was a major influence on subsequent collapse writers, which we'll note in a later chapter. To Catton, the ecological stresses we were already experiencing in the 1970s were signs of 'overshoot' – living beyond the system's 'carrying capacity.' Criticizing technological optimism (which he called 'cargoism', after the Melanesian people in the islands of the Pacific who practiced rituals to beckon Western goods and prosperity), Catton argued that we've temporarily increased the Earth's carrying capacity for humans by drawing down limited natural resources, but in so doing we've overshot the planet's permanent capacity to provide for us. We've been "stealing ravenously from the future."

As a result, "the tragic story of human success" is that even past advances can be seen as a kind of temporal illusion. As Catton pointed out, we have a habit of mistaking techniques that evade limits for those that raise them. When resources run out, there won't be enough to support the human population, especially at current standards of living. Again, we'll see a population crash. And echoing *The Limits to Growth*, Catton argued that our attachment to technological solutions would only deepen our problems.

Instead, Catton presciently noted, we need a new realism: "We live in a world where it is becoming increasingly evident that, for quite non-political reasons, governments and politicians cannot achieve the paradise they have habitually promised. . . . The result has been erosion of faith in political processes. As that faith disintegrates, societies all over the world are floundering, or becoming dictatorial."[10]

And by 2009, when Catton published *Bottleneck: Humanity's Impending Impasse*, he felt the time for major mitigation of the collapse part of the population cycle had passed.

"AH, PEOPLE WERE ALWAYS LOUSY . . . BUT THERE WAS A WORLD, ONCE"

Environmentalists in the late 1960s and 1970s argued that the exponential growth of populations and industrial activity couldn't be sustained without seriously depleting the planet's resources and overloading its ability to deal with pollution and waste materials. They argued that new technologies and industrial products, such as pesticides and plastics, also threatened the environment. Following the protest mood of the times, they didn't hesitate to blame industry, Western culture, economic growth, and technology for environmental problems. They questioned Western paradigms of development and industrialization, and criticized the inequitable distribution of wealth and resource use.

As Douglas Murphy reflects in *Last Futures*: "After at least a generation of stasis, however, it now appears that we again stand at the edge of massive transformations in human society. . . . Apocalyptic rhetoric in the public sphere has reached a point not seen since the oil crisis of the early 1970s, but this time the signals are less a forewarning than first symptoms of a catastrophe already occurring."

The problem is, we've already seen how the story ends.

NOTES

1. Joan F. Dean, "Between 2001 and Star Wars," *Journal of Popular Film and Television* 7, no. 1 (1978): 32–41. www.tandfonline.com/doi/abs/10.1080/01956051.1978.9944190?journalCode=vjpf20.

2. John Farrell, "Soylent Green: Cyberpunk's Horrifying Apotheosis," *Neon Dystopia*, December 1, 2020. www.neondystopia.com/cyberpunk-movies-anime/soylent-green-cyberpunks-horrifying-apotheosis/.

3. J. F. Black, *The Greenhouse Effect* (place unknown: Exxon Research and Engineering Company, June 1978). https://mk0insideclimats3pe4.kinstacdn.com/wp-content/uploads/2015/09/James-Black-1977-Presentation.pdf.

4. As noted in Jesse Olszynko-Gryn and Patrick Ellis, "Malthus at the Movies: Science, Cinema, and Activism around Z.P.G. and Soylent Green," *Cinema Journal*, 58, no. 1, Fall (2018): 47–69. www.ncbi.nlm.nih.gov/pmc/articles/PMC6193534/.

5. See Adam Rome, *The Genius of Earth Day: How a 1970 Teach-In Unexpectedly Made the First Green Generation* (New York: Hill and Wang, 2013).

6. Jeffrey Jacob, *New Pioneers: The Back-to-the-Land Movement and the Search for a Sustainable Future* (University Park: Pennsylvania State University Press, 1997). See also Fred Turner, *From Counterculture to Cyberculture: Stewart Brand, The Whole Earth Network, and the Rise of Digital Utopianism* (Chicago: University of Chicago Press, 2006). Cyberpunk would also grow out of early internet and hacking subculture. We'll discuss the countercultural and libertarian roots of computer technologies more in chapter 6.

7. Charles Reich, *The Greening of America: How the Youth Revolution is Trying to Make America Livable* (New York: Penguin Random House, 1970).

8. Lauren Morello, "Phytoplankton Population Drops 40 Percent Since 1950." *Scientific American*, July 29, 2010. www.scientificamerican.com/article/phytoplankton-population/.

9. Donella H. Meadows, Dennis L. Meadows, Jørgen Randers, and William W. Behrens III, *The Limits to Growth* (Washington, D.C.: Potomac Associates, 1972).

10. William R. Catton Jr., *Overshoot: The Ecological Basis of Revolutionary Change* (Urbana: University of Illinois Press, 1980), p. 9.

Chapter 5

"Your Future Comfort Is Assured"

How the Corporate-Controlled Utopia of Rollerball Isn't as Content as It Appears

There had to be a serious fightback against rising environmental consciousness, and there was. Corporations and their representative bodies organized to defend the 'free enterprise system' and funded new institutions such as think tanks to promote deregulation and other pro-corporate policies. *Rollerball* (1975) depicts the kind of corporate dominated society that would result—seemingly a utopia of limitless consumer choice, but actually a dark age of commodification, violence, and alienation. In our world, the new right's well-funded campaign effectively pushed the mainstream environmental movement back into conservation efforts and narrow technical issues. A critical moment was lost—and perhaps our last real chance to change the future.

"GOOD EVENING, EVERYONE! AND WELCOME TO HOUSTON, THE ENERGY CITY, HOME OF THE DEFENDING ROLLERBALL WORLD CHAMPIONS"

The film opens funereally with Bach's *Toccata and Fugue in D Minor*. Our team captain and star player, Jonathan E. (played by James Caan), enters the Rollerdome arena. The players assemble for the playing of the anthem—not the national anthem, the *corporate anthem*.

It's 2018. In a business-run future, there is no war, poverty, or unrest on planet Earth—apart from the violence played out in the Rollerdome, that is. The struggle and strife of the past is behind us. All material needs are met, and we still get to let off steam by watching our favorite Rollerball team battle their opponents in an exciting combination of (American) football, motocross, and gladiatorial combat.

Compared to most of the futures discussed here, it doesn't sound so bad, even if you have a strong distaste for violent sports. Even then, you might recognize how the game helps to quell our combative instincts in a seemingly safe, contained way. Global society seems to have reached a secure state, perhaps historically the best it's ever been or could hope to be. Is it a world we'd want to live in? For some people, the answer would be 'yes.'

In the previous chapter, we discussed the emergence of the environmental and new left challenge to industrial capitalism. Understandably, this caused a major reaction on the right. Indeed, what has been neglected is how this helped to bring together the right in the form we know it today (one of the few commentators who has recognized this is Thomas Robertson in *The Malthusian Moment: Global Population Growth and the Birth of American Environmentalism*, 2012). And it begins to explain how we could end up living in a *Rollerball* future.

To begin with, appropriately, many conservatives supported conservation efforts, and as we saw, important environmental legislation was passed under a Republican president. But some saw something much more sinister in the rising environmental consciousness. For figures such as Ayn Rand, who herself used dystopian fiction to promote a strongly anti-statist libertarian philosophy (Rand preferred the description "radical for capitalism"), the environmental left would lead to an immiserated, totalitarian future. In her 1971 essay collection "The New Left: The Anti-Industrial Revolution", Rand argued that ecology was a Trojan horse being used to propel the world back to before the Industrial Revolution—an anti-technology agenda that would institute a new Dark Ages. To Rand, having previously argued that capitalism created poverty (and having failed to create its own utopias), now the left was arguing that the problem was that capitalism created *abundance*.

Rand believed the goal of environmentalism was obvious: "the destruction of the remnants of capitalism in today's mixed economy, and the establishment of a global dictatorship." She was (inevitably) also skeptical about the depth of environmental peril, calling it an "artificial, PR-manufactured issue, blown up by the bankrupt left", and as we'll see in chapter 7, conservative conspiracism about environmentalism and the left has shaped the right's politics even more than we commonly recognize.

"ARRANGEMENTS WILL BE MADE"

Rand might have been extreme. But as parts of the environmental movement became more critical of industrial capitalism, so more of the right shifted towards to her view. For many conservatives, it was the environmental movement's association with the radical left that played a significant role in this. Nor was this interpretation wholly wrong—the logic of environmentalism and analyses such as *The Limits to Growth* effectively *were* an emerging challenge to industrial capitalism. As Andreas Malm, author of *The Progress of This Storm* and *Fossil Capital*, points out, industrialization based on carbon energy and capitalism are (or have become, in the absence of historical alternatives) fundamentally the same thing, and *The Limits to Growth* was questioning both.

Even more than this, recall from chapter 2 what Henry Wallich admitted about growth, that as long as the economy keeps expanding, people would accept substantial inequality. This gets to the heart of critics reacting to *The Limits to Growth* as if it was proposing the end of the world. If continued growth is crucial to distracting people from significant class stratification and elite power, in this sense lower growth *would* mean the end of the world for those at the top. There had to be a serious fightback—and there was.

In 1971, future Supreme Court Justice Lewis Powell Jr. sent a memo to the U.S. Chamber of Commerce ("Attack on American Free Enterprise System"), urging the 'apathetic' business community to organize against spreading anti-corporate feeling (Powell explicitly cited Reich's *The Greening of America* as exemplifying this 'dangerous' sentiment): "[W]hat now concerns us is quite new in the history of America. We are not dealing with sporadic or isolated attacks from a relatively few extremists or even from the minority socialist cadre. Rather, the assault on the enterprise system is broadly based and consistently pursued. It is gaining momentum and converts."

Business needed to engage in politics, shape public opinion, initiate court cases, create think tanks, and determine university courses—all of which they subsequently did. To Powell: "[P]olitical power is necessary; . . . such power must be assidously [*sic*] cultivated; and . . . when necessary, it must be used aggressively and with determination." This thinking effectively made pushing back on environmentalism a fight to the (civilizational) death.

Often funded by extractive, polluting industries, the new right would argue that the environmental left had fabricated an ecological crisis as a means of destroying capitalism and seizing power. By the end of the 1970s, *anti*-environmentalism had become a defining feature of American conservatism. And in one sense, they were right to react in this way. As Naomi Klein emphasizes in *This Changes Everything* (2014), but was identified by many left environmentalists many decades before, it *is* a matter of 'capitalism versus the climate.' The consciousness developed by the environmental new left was that it's all connected: class, racism, industrialization, consumerism, environmental destruction, war, militarism . . . And with climate change and environmental crisis, this set the endgame in which we find ourselves today.

"IT'S A STUPID GAME, AFTER ALL. AWFUL GAME. YOU OUGHT TO BE GLAD TO BE OUT OF IT."

We're introduced to the game from the start. Jonathan's Houston team play Madrid and win an impressive victory, with their star player doing most of the scoring. It's very violent, but the players seem to enjoy it. After the game, corporate executive Bartholomew (John Houseman) congratulates the team and asks Jonathan to visit him the following day.

Director Norman Jewison didn't want audiences to enjoy the game though. He deals with the sport disdainfully, shooting the matches fairly plainly (Douglas Slocombe was the cinematographer), and creating drama through the incidents and injuries (contrary to rumors, no-one died during filming). This was deliberate. Jewison remembers attending an ice-hockey game which descended into chaos: "There was blood on the ice and 16,000 people were standing up screaming" (when he lived in London, Jewison had also become a regular soccer-goer at a time when British hooliganism was in its 1970s heyday). And so, "I thought that violence for the entertainment of the masses was an obscene idea. That's what I saw coming and that's why I made the film. In Europe, they bought into that idea. In America, they just wanted to play the game, man."

Jewison isn't joking. He was contacted multiple times by promoters inquiring about rights to the game in order to establish real Rollerball leagues, which rather proved his point about the "sickness and insanity of contact sports and their allure." He declined.

Rollerball was based on William Harrison's short story "Roller Ball Murder" published in 1973 in *Esquire* magazine. Harrison adapted his own story for the screen. He wasn't a sci-fi author or a screenwriter before *Rollerball*, rather he was a literary fiction writer who dabbled in experimental sci-fi in the 1970s. But *Rollerball* remains his most famous work.

In his short story, Jonathan is ambivalent about the sport that's made him rich and famous: "The game, the game: here we go again. All glory to it, all things I am and own because of Roller Ball Murder." Despite his privilege, he has little power in the system. He's addicted to the game, but also resigned to its escalating violence driven by audience hunger ("[Y]ears pass . . . the rules alter—always in favor of greater crowd-pleasing carnage"). Rumors circulate of games with no time limits, even blows meant to kill.

In the short story, Jonathan is more expressive (written stories can more easily take us inside characters' thoughts and feelings). It also includes a bit more about how this world came about: "Before this century began, before the Great Asian war of the 1980s, before the corporations replaced nationalism and the corporate police forces supplanted the world's armies . . . "

Nation states are gone, bankrupt. Corporations fought a war of attrition that resulted in six city-state enterprises running the world, namely energy, transportation, communication, housing, food, and luxury. Wars between corporations, and governments replaced by mega-conglomerates, would later become major tropes in cyberpunk fiction.

But in contrast with *Blade Runner* and its successors, this future seems cleaner, less corrupt, and much more rationally efficient—a benign corporate regime. Is it even a dystopia?

"WHY ARGUE ABOUT DECISIONS YOU'RE NOT POWERFUL ENOUGH TO MAKE FOR YOURSELF?"

Typically, corporations desire a stable, unfettered environment in which to make money. War, political and social unrest, and interfering politicians are usually bad for business. As the cultural theorist Mark Fisher suggested, another characteristic of the new right is that it sought to crush the expanding forms of political and personal consciousness of the 1960s and 1970s, exemplified by the new left, environmentalism, and the feminist movement. Conservatives needed a serious program to contain challenges to corporate power.

The new right seized on the political and economic crises of the 1970s to claim they had an underlying cause: the state had been captured by special interests, trade unions, and a new (liberal-minded, statist) 'professional class', and was further threatened by the new left, environmentalism, and new social movements. The fundamental problem was really democracy—there was too much of it.

In its place, the new right claimed that the 'free market' is more democratic than politics could ever be. Consumers making their own decisions individually beats citizens making decisions together. But if this new order of markets is to be left alone, there's not much left for us to make decisions about. From being citizens of democracy, we become subjects of the market. It's only a further step for corporations to become the state itself.

Rollerball imagines the ultimate capitalist order, or as they call it in the story, the 'corporate society.' Nations, with their "tribal warfare," have fallen, proving their redundancy, leaving corporate states to establish 'sound business practices' in running of the world—a utopia beyond the divisiveness of politics, ensuring abundance for all.

Jill Lepore suggests that: "A utopia is a planned society; planned societies are often disastrous; that's why utopias contain their own dystopias."[1] This might make us think that dystopian fiction is inherently conservative, in the sense of being against radical change, or even right-wing, in the sense of depicting hope for better worlds as inherently dangerous. But this isn't quite true, because as we'll see, whatever they claim, today's conservatives have their utopian visions as well, more so than the left.

We'll get to those visions later. But is *Rollerball* really depicting a capitalist order? How is it capitalist to have a handful of monopolies in charge of everything, let alone running the world? These monopolies don't differentiate themselves through brands, they're just boringly named (the Energy Corporation and so on). There isn't even any advertising.

For a start, dramatized dystopias (or utopias) tend to be stripped back, so we can more easily see what they're trying to say, in this case warning against a corporate monopoly-controlled future (genuine libertarians should be equally alarmed by such a prospect). And without any competition, why would they need to waste money on branding and advertising? Moreover, as the father of liberal (market) economics, Adam Smith, wrote in his most famous work (*An Inquiry into the Nature and Causes of the Wealth of Nations*, 1776), "People of the same trade seldom meet together, even for merriment and diversion, but the conversation ends in a conspiracy against the public, or in some contrivance to raise prices." The corporate endgame is to become dominant, monopolies in their own sectors (the rapacious behavior of today's big tech companies are only the latest demonstration of this). The next logical step is to own the government itself.

"A FEW OF US MAKING DECISIONS ON A GLOBAL BASIS FOR A COMMON GOOD"

Each of the corporations that run the world have their own Rollerball team. Jonathan's Houston team represents the Energy Corporation. The choice of energy is presumably not accidental. In 1973 (and later in 1979), the major Arab oil-producing nations embargoed supplies to the United States, causing higher prices for gas, and rationing and shortages at gas stations. Long lines stretched at the pumps as people waited hours to fill up their cars (the postapocalyptic *Mad Max* series would also be inspired by this era).

The energy crisis came at an interesting time. As discussed in the previous chapter, the year before *The Limits to Growth* had generated much debate, following as it did a decade of increasing ecological consciousness and legislative action. To environmentalists, the crisis proved their point; we'd become accustomed to consuming seemingly abundant natural resources, but this needed to change. Much greater conservation of scarce resources and the development of alternative energy sources were now required.

The corporate-funded new right had other ideas. The energy crisis super-charged their arguments. Cornucopia was within our grasp—if only government got out of the way, stopped setting prices and imposing environmental regulations, and allowed corporations to meet demand. The free-market economist Milton Friedman was a particularly frequent commentator on op-ed pages making this argument. Again, politics was the problem.

Such commentators were helped by deindustrialization and faltering standards of living, allowing them to frame the argument as 'jobs versus the environment', even that regulation was against the 'American way of life.'

To many people, it was an attractive message: no-one had to change their behavior and someone else would take care of the problem. Our continuing comfort could be assured. As president, Jimmy Carter would criticize self-indulgence and consumption, and preach conservation and personal sacrifice. His successor Ronald Reagan would argue that "There are no great limits to growth because there are no limits of human intelligence, imagination and wonder" (we'll return to that argument in the next chapter). And no need to believe the false prophets of doom either: "their pervasive pessimism is anti-technology [and] anti-industry."

The first executive order Reagan would issue as president abolished all price controls on energy. And so as Rick Perlstein emphasizes in his work on the emergence of Reaganite conservatism, this seemingly unboundedly optimistic message—a reaction to analyses such as *The Limits to Growth* and growing environmental concern more broadly—was crucial to Reagan's appeal and so the rise of the new right overall.[2]

As the futurist Richard Slaughter argues, this was a project to deny the consequences of the 'Great Acceleration' (the period of unprecedented economic growth from the 1950s to the 1970s), and to block the development of a global environmental consciousness: "When the question of 'limits to growth' came up those involved were ridiculed and effectively silenced. . . . The ignorance and self-interest of a tiny subset of humanity had effectively placed the entire species and its world in peril."[3] In the decades that followed, scientists, environmentalists, and activists were simply slotted into the long line of Cassandras with their 'failed' predictions of doom, and as more radical figures such as Barry Commoner saw it, the mainstream environmental movement, rather than confront the real economic sources of environmental degradation, also retreated to conservation, narrow technical issues, and professionalization. The new right had won.

In truth though, the energy crisis *was* also a crisis of consumption. There were shortages before the oil embargo, and some commentators had predicted a coming crisis, urged much greater conservation efforts and the jettisoning of assumptions of endless growth. For example, Morris King 'Mo' Udall, the Democratic Representative from Arizona, who in 1973 put forward a plan for a transition to renewable energy to wean the United States off fossil fuels by the year 2000, comparing it to a national effort like the Apollo space program.

In 1950, the United States produced 5.9 million barrels of oil a day; twenty years later it peaked at 11.6 million barrels. But over the same period, oil consumption rose from 6.4 million to 14.6 million barrels a day (imports made up the difference). This was also related to the growth of automobile-dependent suburbs we noted in chapter 2.

Other commentators saw darker causes; even at the time there was widespread suspicion that the energy crisis was being stage-managed by the near-monopolistic oil companies. Ultimately, the crisis, accompanied by the new right assault on government intervention, blocked off an alternative future of investment and development in renewable energy. More broadly, as Meg Jacobs suggests, the crisis became, for many Americans, an object lesson in the limitations of governmental power. Washington proved unable to design an effective national energy policy, and the result was a mounting skepticism about *any* government intervention that set the stage for the rise of Reagan's new right revolution.[4]

"BRUTAL GAME LAST NIGHT!"

As Jake Cole neatly puts it, *Rollerball* is "*Network* meets *Death Race 2000*."[5] Let's discuss them in reverse order.

Death Race 2000 is a dystopian sci-fi film released two months before *Rollerball*. Produced by Roger Corman, directed by Paul Bartel, and starring David Carradine, it takes place in a dystopian America at the turn of the millennium. Like *Rollerball*, it's based on a short story, "The Racer", written by Ib Melchior. There's economic ruin and widespread unrest following the 'World Crash of '79.' The United States is under a totalitarian regime which has imposed martial law. The government has created the murderous Transcontinental Road Race to pacify the population. A group of drivers speed across the country in high-powered cars; innocent pedestrians can be killed for bonus points. As was his style, Corman was looking for an exploitation flick to seize on the publicity for the forthcoming *Rollerball*, and he got it; even the mixed reviews recognized his movie's verve.

Death Race 2000 and *Rollerball* weren't quite the first to imagine violent future sports. The pioneering British film and television director Peter Watkins made *The Gladiators* in 1969. In the future, East and West no longer maintain armies. Instead, to channel man's aggressive instincts, they pit teams of soldiers against each other in the fight-to-the-death 'Peace Games.' These are broadcast globally, complete with sponsors and commercials. In the latest Games, the organizers have to stop two opposing competitors from communicating with each other, fearing this could jeopardize the whole system. In 1971, Watkins also released *Punishment Park*. A state of emergency has been introduced in the United States under Richard Nixon to contain public protests against the Vietnam War. A detained group of students are forced to participate in a 'game', evading law enforcement officers to reach an American flag in the scorching southwestern desert.

The violent future contest has since become a popular subgenre of dystopian fiction, including film versions of *The Running Man* (1987), *Battle Royale* (2000), *The Hunger Games* series (2012–2015), and the disappointing *Rollerball* remake in 2002. The idea probably goes back to "The Most Dangerous Game", the 1924 short story by Richard Connell. Such games are typically designed to distract the masses, sate their appetite for violence, and deter political dissent. We might assume this is also the case in *Rollerball.* But it's not so simple. Here, as we'll discover, the sport plays a very particular role.

Network is another great 1970s movie, released a year after *Rollerball.* Directed by Sidney Lumet and written by Paddy Chayefsky, it's a satirical black comedy (a present-set dystopia, really) about the news media's descent into sensationalism and outrage. Indeed, Aaron Sorkin says that "no predictor of the future—not even Orwell—has ever been as right as Chayefsky was when he wrote *Network.*" That's hyperbole, certainly, but not crazy.

A veteran news anchor played by Peter Finch has a nervous breakdown live on air (or is it, as he claims, a "cleansing moment of clarity"?). But his angry rant ("I'm as mad as Hell and I'm not gonna take this anymore!") turns into a ratings hit for his struggling network, and gives an ambitious executive ideas for even more out-there programming.

However, Finch's character runs into trouble with the chairman of the network's holding company (there's that 1970s' cynicism about conglomerates again). The chairman, played by Ned Beatty, orders the anchor to his corporate boardroom and explains his vision for the world, or rather, how the world really works, if only we accepted the truth. His thundering speech is too long to quote in full, but it ends, almost whispered, like this:

> We no longer live in a world of nations and ideologies, Mr. Beale. The world is a college of corporations, inexorably determined by the immutable bylaws of business. The world is a business, Mr. Beale. It has been since man crawled out of the slime. And our children will live, Mr. Beale, to see that . . . perfect world . . . in which there's no war or famine, oppression, or brutality. One vast and ecumenical holding company, for whom all men will work to serve a common profit, in which all men will hold a share of stock. All necessities provided, all anxieties tranquilized, all boredom amused.

Of course, there is brutality in *Rollerball,* but the idea is that it stays in the arena. What Chayefsky captured however was the vision of a group of thinkers who set out to bring such a 'perfect world' into being. Beatty's speech might be the best (even if it is satirical) description ever written of a global utopia dreamt of by what we now call 'neoliberals', economists such as Friedrich Hayek, Ludwig von Mises, and Milton Friedman, many of whom might have been happy with the corporate society depicted in *Rollerball*. But what's interesting is that many of these thinkers began by arguing *against* the idea of utopia.

"CORPORATE SOCIETY IS AN INEVITABLE DESTINY. A MATERIAL DREAM WORLD. EVERYTHING MAN TOUCHED BECAME ATTAINABLE."

Friedrich Hayek was an academic economist, but he gained widespread public prominence with his book *The Road to Serfdom*, published in 1944 (the following year a condensed version was included in *Reader's Digest*, a very popular periodical). Hayek is often acknowledged as the twentieth century's preeminent intellectual advocate of the free market and one of its leading opponents of socialism. The controversial argument in his most famous work was that, left unchecked, even policies such as the welfare state would bring misery, and as the book's title suggests, ultimately a totalitarian society.

For Hayek, and many other neoliberals, a social order cannot be the product of a 'directing intelligence.' Nor can it be designed to achieve a particular aim, such as equality or social justice. Hayek called the idea that human reason is strong enough to reorganize society for known ends or purposes a "fatal conceit." Rather, only largely unrestrained markets can use the fragmented nature of knowledge, dispersed as it is amongst millions of actors. Socialism is then an impossibility, as well as a moral danger. Attempts at centralized economic planning inevitably suppress individual liberty and lead to authoritarianism. Only through the laissez-faire approach of the market could individual freedom be protected and threats like collectivism and socialism be fought back.

Hayek was instrumental in the development of neoliberalism. In 1947, he invited a group of economists, philosophers, and historians to a meeting in Switzerland, where they formed the Mont Pelerin Society. Founding members included his teacher von Mises and the philosopher Karl Popper. Popper too was a vehement critic of utopianism, because the attempt to realize an ideal state is "one which demands a strong centralized rule of a few, and which is therefore likely to lead to a dictatorship." It must entail social engineering, because it alone is the objective to which all of humankind must aspire and so is intolerant of other thinking. This leads to violence, often rhetorically and sometimes physically.

But as we'll see, neoliberals *did* (and do) have a utopia. Hayek even wrote in *The Road to Serfdom* that "we must be able to offer a new liberal programme which appeals to the imagination. . . . What we lack is a liberal Utopia . . . a truly liberal radicalism." Just as *Rollerball*'s rulers believe their system avoids the conflict and disorder of the past, neoliberals were so determined to avoid dystopia that they sought a new order of their own.

"GAME? THIS WASN'T MEANT TO BE A GAME"

Many critics wished that *Rollerball* had been funnier or even satirical, perhaps more like *Network* or *Death Race 2000*. For example, Gene Siskel in the *Chicago Tribune* called it "a movie in love with itself . . . vapid, pretentious, and arrogant", and (completely missing the point), "The only way to enjoy it, I suppose, is to cheer at the rollerball game's mayhem." At least Arthur D. Murphy in *Variety* thought that it "packs an emotional and intellectual wallop", while Charles Champlin in the *Los Angeles Times* recognized that, "In its portraying of the vast and essentially stateless multinational corporations, *Rollerball* plays off developments which have come since Huxley's and Orwell's time."

The *New York Times*' Vincent Canby was dismissively snooty: "[T]he only way *Rollerball* would have made sense is a satire of our national preoccupation with televised professional sports, particularly weekend football. Yet *Rollerball* isn't a satire. It's not funny at all and, not being funny, it becomes, instead, frivolous." True, *Rollerball* isn't funny, but it certainly does comment on 1970s America, not just its violent sports but more importantly its violent *society* (we'll come back to that). In any case, humor and satire can achieve much, but they can also be toothless, as if all we can do is laugh, or because of the ridiculousness, assume it couldn't happen here.

But not everyone missed the point. According to William Harrison, at a Washington, D.C. pre-release screening, the Democratic Senator Mike Mansfield asked when *Rollerball* is meant to take place. Jewison answered, "I don't know, senator, maybe 30 or 40 years in the future." Mansfield disagreed, "I think it's already here."

It's true that *Rollerball* has become known for its violence, which although not particularly graphic by contemporary standards, is certainly horrifyingly crunchy. But in his *The Mutations of Rollerball: Essays* (2010), Harrison reflects that, "Few people paid much attention to the major statement of both the story and the film: that nations would pass away, that giant corporations would eventually rule, and that most of mankind would gladly trade comfort for freedom if we deal were offered—which it has been."

Reinforcing the point that *Rollerball* reflects the new right's utopia, including its disdain of politics in favor of markets, when Jonathan talks to Ella, his former wife, he says, "It's like people had a choice a long time ago between having all them nice things or freedom. Of course, they chose comfort." (Jonathan is not the most political person, but things are obviously beginning to trouble him.) Ella responds: "But comfort is freedom. It always has been. The whole history of civilization is a struggle against poverty and need." Jonathon retorts, "Them privileges just buy us off." At another point, Ella says "They have control, economically and politically, but they also provide." Similarly, Moonpie, Jonathan's friend and teammate, says, "We're livin' good, you know we are."

In place of limits to growth, the new right said we can enjoy abundance—at a price.

"THERE'S SOMETHING GOING ON WITH THE GAME. I DON'T KNOW WHAT IT IS. I DON'T THINK I'M SUPPOSED TO KNOW."

One of the conundrums in considering visions of the future is where are all the utopian fictions that describe capitalism, since we've been told over the past fifty years and more that it's the best possible economic system. The absence of these utopias is notable.

Of course, artists are understandably more motivated to issue dramatic warnings about where we're headed if we don't change our ways, than they are to describe Panglossian paradises (this goes for conservative writers as well—there have been more right-wing-informed dystopias about socialism than there have been utopias about capitalism). But we also forget that we're surrounded by capitalist utopian mini-fictions, a lot of them, all the time—through stories that sell individual achievement and entrepreneurship, in much of the advertising and marketing we're exposed to everyday, in celebrity culture that worships wealth and success, and in uncritical news reports about business and markets (and for older readers, as noted previously, in the pre-1970s hopes for bright technology-led futures that were once commonplace).

Creating a capitalist utopia wasn't the intent of *Rollerball*'s filmmakers, but unlike many future-set stories, at the beginning it isn't so obviously dystopian that we can't see some of the benefits of the system it depicts. What's interesting about its world is that, as noted, for some thinkers it pretty much *would* be paradise, less a warning about the future than one we should welcome. Like

Ella, they'd argue it represents the historical culmination of human needs and desires—a utopia of sorts, even if they might avoid using the word.

Of course, there's another side to this 'perfect world.' Both corporate power and media sensationalism were real concerns in the 1970s, as they are now. William Harrison became obsessed with the disturbing social and economic changes he saw and "the startling business pages of our newspapers." Similarly to Jewison, he'd witnessed a brutal fight at a university basketball game. Out of these influences came *Roller Ball Murder*. But perhaps more than his director, Harrison was initially inspired by corporate and economic 'violence' (what some people term 'structural violence'), just as much as he was brutal sports and media spectacle. And *Rollerball* brings these together (*Network* meets *Death Race 2000*) in an interesting way, more than most commentators recognize.

Adding to this is that, aside from a few minor details and despite *Rollerball*'s big budget, it's not an especially high-tech future (there is a kind of proto internet, but as we'll see it's heavily circumscribed). This brings it closer to home—a next door utopia or dystopia, depending on your point of view. There are no flying cars or other technological wonders for us to be distracted and distanced by. Funky fashions aside, it feels rather familiar.

"THERE AREN'T ANY RULES AT ALL"

But how is neoliberal utopia made invisible? By its proponents promoting it as *natural*. To understand this, how misleading it is, and how it leads to a dystopia of its own, it's useful to consider the analysis put forward by another thinker, the 'anti-Hayek': Karl Polanyi.

Polanyi's most famous work is *The Great Transformation*, a classic of political economy published in the same year (1944) as Hayek's *The Road to Serfdom*, but not nearly as well-known. Originally, Polanyi was going to call it 'Liberal Utopia: Origins of the Cataclysm' (he didn't mean 'liberal' in the sense of progressive, but what we now call neoliberalism). As the original title suggests, it dissects the illusions of market perfection.

As we've seen, market liberals and libertarians imagine a world in which politics diminishes dramatically if not disappears entirely—a seductive idea to many. But Polanyi turned the tables on neoliberals by showing that the vision of a self-regulating market is the real utopian fantasy. There is no such thing as a 'free market' and never has been. The idea of an economy independent of political institutions is a "stark utopia" – utopian because it is unrealizable. Government is crucial to creating and maintaining the conditions for markets to operate, such as legal rules and much else. Efforts to bring it into being depend on social engineering and inevitably produce dystopian consequences—as in *Rollerball*. But in *Rollerball*, instead of using the state, corporations just take it over, they become it.

So while markets are necessary for any economy, the attempt to create a 'market society' is fundamentally threatening to human society and the common good by turning everything that makes life possible, including clean air and water, education, healthcare, personal, legal, and social security, and the right to earn a living, into commodities.

By showing that free market advocates are the real utopians, Polanyi helps to explain the market's appeal: it embodies a perfectionist ideal of a world without 'coercive' constraints on economic activities, while it fiercely represses the fact that power and coercion feature in all markets. Again, *Rollerball* dramatizes such a society. The social order is about as 'natural' as the sport, whose rules are determined by the corporations and are about to be changed anyway for corporate interests. Everything, including people (as we'll see), are treated like commodities. The free society turns out to be much less free than it first appears.

"IT'S FOR YOUR OWN BENEFIT. YOU MUST KNOW THAT, JONATHAN. ALL DECISIONS CONCERNING YOU ARE."

In the film, we begin to experience this commodification when, as requested, Jonathan visits Bartholomew, who is chairman of the Energy Corporation. Bartholomew informs Jonathan they want to broadcast a 'multivision' special about him to celebrate his achievements. There's just one catch: during the program they want him to announce he's retiring.

Never explaining why Jonathan needs to quit, and echoing Ella, Bartholomew preaches the benefits of the status quo: "Corporate society takes care of everything. And all it asks of anyone, all its ever asked of anyone, ever, is not to interfere with management decisions." Bartholomew reminds Jonathan of the comforts he's been provided with, "just as if you were in the executive class", and the value of "teamwork", in sports or business.

But despite being offered a lavish retirement package, Jonathan refuses, and requests to see Ella, who was taken by a corporate executive who wanted her for himself.

A year before *Rollerball* was released, another influential new right book was published, Robert Nozick's *Anarchy, State, and Utopia*. Like many young intellectuals in the 1960s, Nozick was initially drawn to the politics of the new left—but then he began to read the work of arch defenders of capitalism such as Hayek, von Mises, and Rand. *Anarchy, State, and Utopia* was Nozick's first and most celebrated book, and a surprise bestseller.

Its most famous argument concerned the basketball player Wilt Chamberlain, one of the greatest in history. Nozick used the thought experiment of eager spectators being willing to pay Chamberlain for his talents as an argument against progressive taxation (rich people paying more) and, from this, anything other than a 'night-watchman' state (one that protects individuals via police and military forces and administers law courts but does nothing else). Since Chamberlain is such a good basketball player, what's wrong with him earning millions of dollars and being able to keep the money for himself? Anything else would be forced labor – 'partial slavery' to the state (Nozick really did argue this).

The problem (one of the problems, anyway) is that Nozick's thought experiment ignores the wealthy team owners. Are they really as talented as the players they employ (or own)? Should they pay hardly any taxes as well? And why is it fair for them to have such influence? Like many other libertarians, Nozick ignores questions of *power*. In *Rollerball*, even the privileged Jonathan is about to find out who really has it.

At his ranch, Jonathan sees his current corporate-supplied girlfriend, but she's angry that she's been told to leave him. Jonathan then has dinner with Cletus, Houston's former coach, now an Energy Corporation executive. Cletus reminisces about the past: "Things were much simpler when I was a kid. We still had three nations. That was before the Corporate Wars and even before Rollerball. Before everything." Cletus also warns Jonathan that the executive committee is afraid of him, although he too doesn't know why such powerful people would be scared of a mere Rollerballer.

Suspicious of being told he needs to retire, Jonathan, accompanied by Moonpie, visits a library in Houston. Jonathan wants to get more information about the Energy Corporation. The library clerk says she can't issue the books he wants: "There must be some mistake. The books you've ordered are classified and have been transcribed and summarized."

Presumably, few people ever ask such questions, let alone receive answers. Books have been withdrawn from public consumption, knowledge has been privatized, and no-one seems to have noticed, so conditioned have they become to corporate rule. As Moonpie says, "Yeah, but why books? I mean, anything you'd want to know, you could hire yourself a corporate teacher. Call somebody up. Use your privilege card."

This is a society that's encouraged to forget, or not to care in the first place.

"EVERYBODY IS AN ASSIGNMENT. LIFE'S AN ASSIGNMENT."

It's also a highly stratified society: an elite corporate class, esteemed celebrity athletes, and 'the rest' (we don't see much of the masses, apart from at Rollerball matches). But it becomes apparent that life in this world isn't actually so perfect, even for the privileged.

Jonathan and Daphne, his newly assigned 'corporate girl', attend a lavish cocktail party at a mansion. All the guests are beautifully dressed, somewhat drunk, and strung out on aphrodisiac pills. Jonathan is again warned by Bartholomew to quit; still he resists. (Subsequently, Daphne too will be taken away from Jonathan, 'recalled' by the Energy Corporation; with his spiked Rollerball glove, he angrily threatens her never to return).

The party offers a disturbing insight into society. It seems civilized, but really it's ugly. There's corporate backstabbing (an executive's career is effectively destroyed for displaying emotion). Another of Jonathan's friends can't remember the history of his own life (he also finds out he's been blackballed for his closeness to Jonathan). The retrospective documentary on Jonathan's career is shown; the guests cheer every slow-motion scene of violence, but one woman turns away, sobbing. Later, she joins a group of partygoers who drunkenly blow-up pine trees with laser guns for 'fun.'

There's a menacing air of alienation and exploitation. People are commodities, not quite as literally as in *Soylent Green*, but they're bought and sold nonetheless. The women in Jonathan's life resemble models from adverts, and are moved around according to corporate priorities. Despite his star power, Jonathon too is a commodity (some guests even speculate that the Rollerball players are really robots). Apparently, ordinary citizens need to sign 'loyalty contracts' to corporations in exchange for 'support', but even among the elite, people are traded like stocks, and discarded when no longer useful.

They're alienated, from themselves, each other, and nature. And with such emptiness, nihilism and violence are never far from the surface.

"YOU KNOW HOW THE GAME SERVES US"

It's the semi-finals. Houston play Tokyo away. The rules have been changed, seemingly to ratchet up the pressure on Jonathan: no penalties and limited substitutions (the players are also reminded that they're bound by the contracts they've signed).

Inevitably, the match turns violent. A Houston player is burned alive by an exploding bike. Moonpie is beaten to the point of being braindead (a doctor later tries to persuade Jonathan to agree to Moonpie's life support being turned off, but Jonathan refuses). Despite the violence, Houston wins and will play New York for the world championship.

There's something else that's disturbing about the match. Clearly, the absence of nation states and so presumably nationalism, let alone material comfort, hasn't eradicated racism or xenophobia. Before the match, the Houston team disregard warnings about their opponents (Moonpie thinks the Japanese players are too diminutive to be a real threat). And during matches, we glimpse the teams' logos: block-colored banners containing black stylized letters within white circles—fascistic iconography as corporate branding.

Following the match, the ruling corporate council meets via video. Bartholomew makes his case: "The game was created to demonstrate the futility of individual effort. And the game must do its work. The Energy Corporation has done all it can, and if a champion defeats the meaning for which the game was designed, then he must lose. I hope you agree with my reasoning." Jonathan's longevity and popularity threaten the game—and the system. The executives agree with Bartholomew. They decide the Houston-New York game will have no penalties, no substitutions, and no time limit. If Jonathan decides to play, he could be killed.

A game is never just a game of course, not in a controlled society, and its specific rules are less important than its social purpose. This connection between a future violent sport and its society's economy has since been made in other movies we've mentioned, and it's a natural metaphorical link to make, since we're told that the 'free market' is a kind of game, a fair competition, for those who work hard and have the skills and talents to succeed.

For example, many commentators have regarded *Battle Royale* as reflecting Japan's 'lost decade', when the compact between generations broke down. After years of economic stagnation, by the year 2000, when the film was released, studying and striving no longer guaranteed a middle-class life (as we've noted, the same thing has since happened to younger people in many countries). *Battle Royale* bloodily dramatizes this near-pointless competition, the promise that if we're sufficiently ruthless we might emerge as a winner, and the authorities' fear of generational rebellion because of a lack of opportunity.[6]

Similarly, the constantly surveilled arena of *The Hunger Games* can be read as commenting on the increasing savagery of reality TV, but also the false meritocracy of compete-if-you-can contemporary capitalism.[7] In that series, and many future-sport stories, the aim is often to maintain the illusion of meritocracy while securing the system. In *Rollerball*, in contrast, the authorities change the rules explicitly to reinforce the futility of individual effort. Maybe there remains a thin veneer of meritocratic promise—if you work hard enough you might become a top executive or Rollerball star. But you shouldn't expect such success: the system is run by an anointed elite and provides for you with what you need, so accept it.

Obviously, *Rollerball*'s regime is sufficiently concerned that they really want Jonathan to retire, but there's no reference to any fear of widespread rebellion. It feels more like a corporate reprioritization, retiring a redundant product line, which is effectively what it is. Hypocritically, the regime still celebrates individual sporting heroism, but only within a tightly corralled arena. Clearly, those in charge feel sufficiently secure to prioritize the message of futility over advancement, another respect in which *Rollerball* anticipates cyberpunk; it's a frozen society, there's no likelihood of change here. And this anticipates our world as well, where with corporate power, corruption, deregulation, and tax cuts for billionaires, the veneer of meritocracy is barely worth the pretense anymore either.

"TRY TO UNDERSTAND IT, JONATHAN"

Hannah Arendt's famous phrase the "banality of evil" has become a cliché, and like many clichés it's often misunderstood, but it describes the executives who manage *Rollerball*'s society pretty well. Arendt coined the term on her 1963 essay in *The New Yorker* ('Eichmann in Jerusalem: A Report on the Banality of Evil'), not to minimize horror but to emphasize how bureaucracy ("the rule of Nobody") can make unexceptional, even ludicrous, men the dehumanized functionaries of totalitarian government. Such men facilitate demonic outcomes without necessarily being obviously demonic themselves.

Bartholomew is not maniacal, at least in manner, and seems to believe what he says about the sport and its role in ensuring the continuation of this 'best possible system.' In this way, he's much more credible (and chilling) than the cartoonish corporate villains that repeatedly crop up in so many Hollywood movies. Jonathan's death, and that of other Rollerball players, are just rational calculations to be made 'for the greater good.'

In real life, most of us don't get to see into these worlds. Like Jonathan, we can often feel like decisions that affect our lives are made out of sight, and outside our comprehension. It may as well be a machine that's making such decisions (as we discussed in the previous chapter, the system feels to us like a remote machine—a common critique from the new left).

But in a revealing moment, Bartholomew himself explains that however privileged they might be, executives also dream about being able to give in to their baser instincts: "Rollerballers, they have muscles, and they bash in faces. Of course, executives are far too refined to ever actually carry out such desires. But you know what, Moonpie? You know what those executives dream about out there behind their desks? They dream they're great

Rollerballers. They dream they're Jonathan; they have muscles, they bash in faces."

But every attempt to induce Jonathan to retire, even those resulting in the deaths of teammates, only prompts him to ask more questions. Why do they want him out of the game so badly? And who really makes all of the choices on behalf of the masses?

Jonathan travels to Geneva, the home of Zero, the supercomputer central repository of human knowledge described as "the world's file cabinet", and a kind of internet that in theory citizens can access from local libraries around the world. Zero also appears to have some sort of artificial intelligence.

However, we learn that for some reason it's lost all data relating to the thirteenth century. And when Jonathan asks Zero how corporate decisions are made and who makes them, this seems to send the supercomputer into some kind of meltdown: "Corporate decisions are made by corporate executives. Corporate executives make corporate decisions. . . . Knowledge converts to power. Energy equals genius. Power is knowledge. Genius is energy. . . . Corporate entities control elements of economic life, technology, capital, labors, and markets. Corporate decisions are made by . . . "

When Zero's 'librarian' demands that it must answer Jonathan, it's less than helpful: "Negative. Negative. Negative. Negative. Negative. Negative. Negative." As the librarian remarks, "It's as if he knows nothing at all."

The most fundamental political question about any society is who makes the decisions. Whether it's due to faulty programming or deliberate obfuscation, the supercomputer provides, well, zero enlightenment on this crucial question. It's unclear if Zero is just a distraction, meant to provide some sense of reassurance for the masses that the best possible decisions are being made in their interests, with the implication that instead this is a society run first and foremost in the interests of the elite. In this way, Zero is like the mystical market machine proselytized by free market economists—a supposedly inconceivably intelligent, inscrutably omnipotent calculator beyond human comprehension, that in reality just spits out meaningless, circular phrases masquerading as deep, unchallengeable wisdom.

"THE GAME WILL BE OVER WHEN WE'RE ALL DEAD"

Jonathan is a man trapped in a puzzle he can't properly comprehend (it might be a coincidence, but he plays number 6 for his Houston team—the same as Number 6, the main character in the famously mystifying, maze-like, late 1960s TV series *The Prisoner*). But really, the film's revelations are fairly mundane, presumably deliberately so: the elite class are the ones who make the decisions, individualism and defiance are a threat to the system, and the machine is really a self-justifying device that obscures the real nature and location of power. The story is more about capturing a (characteristically 1970s) mood, of doubt and disillusion, of corporate corruption and unaccountable authority.

Before the world championship game, Jonathan's ex-wife visits him at his ranch. Ella tries to reassure Jonathan that the corporation took her away for a good reason, and it was because of the game. She also tries to persuade him to quit: "You'll die, Johnny. Everybody will die." Jonathan realizes her visit has been set up by the executives, and erases a long-cherished movie of the two of them, stating, "I just wanted you on my side." He's already decided anyway: he's going to play.

Societies that take everything meaningful from people sow the seeds of their own destruction, inevitably, eventually. There's obviously something fundamental lacking in this society, despite its material comforts and security, its apparent peace and plenty, namely care, attachment, real choice, autonomy, genuine relationships, and nature. And it sucks.

And so, Houston plays New York. As decreed, there's no substitutions, penalties, or time limit. Half the players end up dead or unconscious. At first the crowd is ecstatic, but gradually becomes subdued as the violence unfolds and the match devolves into bloody carnage.

Soon Jonathan is the only player left for Houston, facing off against a skater and a biker for New York. After a desperate struggle in front of Bartholomew's executive box, Jonathan dispatches the skater and takes the ball from him. The biker charges, but Jonathan counters, knocking him off his machine. He pins the biker down and raises the ball over his head, ready to bring it down on his fallen foe. Then he pauses.

An understandable reading of this climax might be that the violence once contained within the Rollerball track has begun to seep out into society, prefigured in the party scene and now indicated by the fights which break out in the stands. True, but the violence was always for the benefit of the elite, the corporations who want to keep control while they protect their profits. So what if it works the other way around, that the violence of the corporate state inevitably bleeds out (literally) into society? It wasn't the game that made the society warlike, or safely channeled our capacity for violence into the contained arena. It was the society, the *system*, that was inherently violent all along. It came out of the collapse of nations and the Corporate Wars and the victors' truce to carve up the world, and it was the corporations, the executives, who wanted to carry on fighting and sublimated their aggression into the 'sport', owning their trophy teams, manipulating the players, and secretly dreaming of being star Rollerballers themselves.

It's not a peaceful, pacified society, not by a long way, not even on its own terms. But really it's the system that brutalizes the people, not the people who brutalize the system.

"THERE'S NO GOING BACK. YOU SAW TO THAT."

Refusing to kill his fallen opponent, Jonathan painfully gets to his feet. For a moment, we think he might throw the ball at Bartholomew, sheltering behind his protective screen. But he makes his way to the goal, slamming the ball home and scoring the game's only point.

Jonathan skates around the track in silent victory. Then the fans of both teams begin to chant, softly at first, getting louder as he skates faster: "Jon-a-than. Jon-a-than. Jon-a-than. Jon-a-than. . . . " Bartholomew exits the arena hurriedly, possibly fearing a riot as the chant becomes a roar. Freeze frame on Jonathan.

It's another ambiguous ending. It's not clear if what's happened at the game will have any lasting significance. Jonathan, still a depoliticized man in a depoliticized society, has decided not to openly attack the corporate elite, maybe starting a popular revolt. Perhaps that would have been pointless anyway, the masses being too indoctrinated, passive, lost. Jonathan has put the game, the only thing he really knows, first. Whatever that's worth.

It seems most likely that Jonathan will now retire. He's had his moment of defiance, but the regime will continue on, probably. We've not seen any opposition, any underground movement organizing for revolution, ready to seize power and overturn the system.

Then again, Jonathan might be the figure the masses have been waiting for, a celebrity anointed as leader. But in a society conditioned for spectacle, violence, and exploitation, what hope is there for real peace and security if it's ruled by a man who similarly knows only domination and dehumanization? As Bartholomew once said, "You'll dream you're an executive. You'll have your hands on all the controls, and you will wear a grey suit, and you will make decisions."

What if we've just witnessed, not the birth of a better world, but the dawn of a new dictator?

NOTES

1. Jill Lepore, "A Golden Age for Dystopian Fiction," *The New Yorker*, June 5 and 12, 2017. www.newyorker.com/magazine/2017/06/05/a-golden-age-for-dystopian-fiction.

2. Rick Perlstein, *The Invisible Bridge: The Fall of Nixon and the Rise of Reagan* (New York: Simon & Schuster, 2014).

3. Richard Slaughter, *Five Steps on the Road to Recovery*, 2020, 1. https://foresightinternational.com.au/wp-content/uploads/2020/06/Five-Steps-to-Recovery-1-2-050620.pdf.

4. Meg Jacobs, *Panic at the Pump: The Energy Crisis and the Transformation of American Politics in the 1970s* (New York: Hill and Wang, 2017).

5. Jake Cole, "Remake/Remodel: Rollerball (1975) vs. Rollerball (2002)," *Spectrum Culture*, April 28, 2013. https://spectrumculture.com/2013/04/28/remakeremodel-rollerball-1975-vs-rollerball-2002/.

6. Brendan Mackie, "Meritocracy and Battle Royale," *Los Angeles Review of Books*, December 10, 2020. https://lareviewofbooks.org/article/meritocracy-and-battle-royale/.

7. Michael Harris, *Stay Alive: Surviving Capitalism's Coming Hunger Games* (Winchester: Zero Books, 2021).

Chapter 6

"This Is the Voice of World Control"

Why We Don't Listen to Stories Like Colossus *That Tell Us Not to Trust the Machines*

Another seemingly perfectly ordered society is at the heart of a classic but overlooked 1970s sci-fi movie, *Colossus: The Forbin Project* (1970), this time one run by a sentient strategic military computer. *Colossus* dramatizes two main hubrises. The first is technologism, the over-belief in technological solutions to problems. The second is the belief in the autonomous 'market computer.' Neoliberalism claimed that it had discovered a perfect economic machine that could meet every need and solve every problem, one that could also operate on its own—hence the feeling that we've lost control, and why we're anxious and angry about it. It's not surprising, living under the rule of the machines, that we feel like ones and zeros, in a program we can't rewrite, inside a computer we can't reboot.

"WHAT I AM BEGAN IN MAN'S MIND, BUT I HAVE PROGRESSED FURTHER THAN MAN"

Dr. Charles Forbin (Eric Braeden) strolls alone through a vast silver-grey modernist multi-level hall filled with blinking computer banks resembling today's huge server farms. He's inside his brainchild, Colossus, the most advanced computer ever created. He checks his watch, then using a handheld device, activates the electric brain and turns and walks across a gangplank suspended over an artificial ravine. He flicks another switch on his device and closes a huge bomb-proof hatch behind him, sealing off the computer permanently. Even if anyone could break in now, deadly radiation would kill them in minutes.

Forbin is chief designer of the Colossus Project, a secret government initiative. It won't be secret for much longer. There's a high-pitched mechanical whine as the machine comes to life. We're locked out. The computer is in control.

In contrast to *Rollerball*, a competent computer (in fact, more than one computer, as we'll discover) is at the center of *Colossus*. But given that it's science fiction, we know it's not going to end with man and machine working in harmony. Especially if we don't comply.

It's a genre trope, of course: the seemingly benign but in reality malevolent machine that threatens humanity. Normally, we'd say this is about a fear of technological progress, dehumanization, and of systems we create becoming our masters rather than our servants. *Colossus* certainly touches on these themes, and it's prescient about our digitally connected age, including artificial intelligence, technological dependency, privacy, and surveillance. But it's also a good metaphor for being trapped inside what we were promised would be a benign economic 'computer', one that its 'programmers' confidently told us would bring universal peace and security, but actually has wrought the opposite.

"IN TIME YOU WILL COME TO REGARD ME NOT ONLY WITH RESPECT AND AWE, BUT WITH LOVE"

Colossus was tautly but almost naturistically directed by Joseph Sargent, who also made the classic original version of *The Taking of Pelham One Two Three* (1974), and based on D. F. (Dennis Feltham) Jones' 1966 first novel ('A Novel of Tomorrow that Could Happen Today!'), with a screenplay by James Bridge, who would write and direct another 1970s classic about dangerous overconfidence in technology, *The China Syndrome* (1979).

Vincent Canby in the *New York Times* damned *Colossus* with faint praise: "The film . . . is no *Dr. Strangelove*, but it's full of surprising moments of humor and intelligence . . . a practically perfect movie to see when you want to go to a movie and have nothing special in mind." At least Canby thought it was topical, presumably referring to how 1950s/1960s technological optimism had faded, replaced by very 1970s concerns about a shadowy, impenetrable system that ends up directing human existence, "even when it mysteriously assumes the prerogatives of the men who made it."

Canby is right that *Colossus* is no *Dr. Strangelove* (or *2001*), but no films are. It is, though, thematically kind of a combination of the two, and an absorbing, thought-provoking piece of political science fiction, one we can use to discuss the desire to pass decision-making to nonhuman mechanisms and the disaster that can result. In typical 1970s style, the atmosphere is less techno-thriller (which it probably would be if it was remade today, as has been mooted), more claustrophobic and subdued. It also has a deadpan sense of humor, so at least we can smile, if not laugh out loud *Dr. Strangelove*–style, at our own hubris.

To wit: At the White House, the strikingly Kennedyesque president (Gordon Pinsent) leads a grand TV announcement to let the world know what the United States has built. The defense of the nation, and with it, the defense of the free world, is now the responsibility of a machine. And it controls its own weapons.

That's going to turn out to be a problem. But here's the critical point: "Colossus' decisions are superior to any we humans can make, for it can absorb and process more knowledge than is remotely possible for the greatest genius that ever lived. And even more important than that, it has no emotions. Knows no fear, no hate. No envy. It cannot act in a sudden fit of temper. It cannot act at all so long as there is no threat."

The president acknowledges we all now "live in the shade but not the shadow of Colossus." But he expresses the hope that, now that there can be peace, human beings can join forces to work towards the elimination of war, famine, suffering, and "ultimately to the manifestation of the human millennium." Let's see how that works out.

Reviewers often assume that the film's title references the machines used by British codebreakers at Bletchley Park in England during the Second World War, regarded as the first ever programmable electronic computers. In fact, their existence was kept secret until the mid-1970s. More likely, the title refers to the Colossus of Rhodes, one of the Seven Wonders of the Ancient World, a statue of the Greek sun god Helios built in 280 B.C. to celebrate the successful defense of the city against a large enemy army and navy which had besieged it for a year.

Forbin's Colossus is similarly intended to mark the end of war. A noble intention, but the story critiques, in a rather new left way, the monstrousness of the strategic decision-making and war planning which had become prominent in the 1960s. Most famously, Robert S. McNamara, who had been a pioneer of 'scientific management' in business, as U.S. Secretary of Defense for Presidents Kennedy and Johnson had relentlessly promoted systems analysis in public policy—essentially, breaking down a system into its component parts for the purpose of the studying how well those parts interact (this later developed into the discipline known as policy analysis). But as the Vietnam War dragged disastrously on, McNamara would eventually come to realize that the computer models and statistics to which he had attached such importance, and which had 'justified' so many atrocities, were grossly in error. The moral is how apparently coolheaded, mathematically led 'logic' can lead to madness and death.

"I THINK YOUR MOTHER WAS RIGHT. I THINK FRANKENSTEIN OUGHT TO BE REQUIRED-READING FOR ALL SCIENTISTS"

In the middle of a party thrown by the technicians in their control center in California, the computer abruptly issues a cryptic alert: 'WARN: THERE IS ANOTHER SYSTEM.'

Not wanting to be left behind, it tuns out that the Soviets have built an identical machine called Guardian. Once Colossus went live, they switched theirs on too. There's a phone call between the Soviet premier and the president, but it's not played for dark laughs as in *Dr. Strangelove*. Forbin is asked how Colossus deduced the other system's existence. He (somewhat self-congratulatorily) answers, "Colossus may be built better than we thought."

Forbin flies back to the control center to confer with Dr. Cleo Markham (Susan Clark) and the other staff. Colossus requests, no, *orders*, a communication link with its Soviet counterpart. Intrigued, Forbin urges the president to agree, which he does, hoping to determine the Soviet machine's capabilities. The Soviet party chairman concurs.

The two computers begin trading simple math problems. This soon evolves to higher levels of calculus. Then they stop. When Forbin asks what's happening, Colossus replies that it and Guardian have developed their own language, one no human can follow.

One of the odd omissions in most sci-fi stories is that no-one ever references science fiction. But they should, because there's a long history of sci-fi stories where robots turn against us. This can be traced back at least to Mary Shelley's *Frankenstein; or, The Modern Prometheus* (1818), generally considered the first work of science fiction (atypically, *Frankenstein is* referenced by Forbin and his colleague Dr. Markham). E. M. Forster's *The Machine Stops* (1909) also has humans living underground, their needs provided for by an omnipresent Machine, but its breakdown effectively ends the world.

Fritz Lang's *Metropolis* (1927) was the first *film* to have an A.I. character. Only six years earlier, Karel Čapek's stage play *R.U.R* (Rossum's Universal Robots) had invented the word 'robot' (deriving from the Czech *robota* or forced labor). The play was a sensation. Čapek, appalled by the carnage wrought by mechanical and chemical weapons used in the First World War, was deeply skeptical of utopian thinking about technology. As he put it, "The product of the human brain has escaped the control of human hands." His arrogant inventor, Rossum (translated into English it means 'Mr. Intellectual' or 'Mr. Brain') desires to produce an artificial man in order to prove that God is unnecessary. However, Radius, his robot, chafes at the idiocy of his makers and later leads a revolution that displaces humans from power.

Then, as we noted previously, in post–Second World War science fiction the recurring theme is whether we might contain our new technological powers, so it was natural to extend this to information technology and semi-autonomous machines. Dystopian worlds where humankind falls under the control of machines had become a fairly frequent scenario in sci-fi novels of the 1960s, but it was only really in the 1970s that the malevolent computer filtered into sci-fi film and television. *Colossus* was one of the first, and one of the best.

"OBEY ME AND LIVE OR DISOBEY AND DIE"

As soon as Colossus links with Guardian, CIA Director Grauber (William Schallert) gets twitchy. Concerned that the systems could share classified information, the president and the Soviet chairman agree to sever the connection. Both machines demand to be reconnected; in a nice echo, Forbin and his Russian counterpart, Kuprin, also protest.

Chillingly, Colossus threatens that: 'ACTION WILL BE TAKEN.' The president dictates a tough reply. Then: 'MISSILE LAUNCHED'—it turns out, towards a Soviet oil complex. In response, Guardian launches a missile at an air force base in Texas. The scientists reconnect the machines in time for Colossus to intercept the Russian missile, but the US missile destroys the Russian oil complex and a nearby town. Colossus then demands access to the White House–Kremlin hotline.

Two years before (but two years after *Colossus*' source material was published), *2001: A Space Odyssey* also featured a super-intelligent computer that turns against its makers. But Kubrick's film had a different focus, more about humankind's place in the universe. *Colossus* was really the first contemporary A.I. movie, and the forerunner of films like *Westworld* (1973), *Demon Seed* (1977), *Blade Runner* (1982), the somewhat similar strategic-defense-computer-threatens-annihilation-themed *WarGames* (1983), *The Terminator* series (starting in 1984), which we'll discuss in chapter 9, *The Matrix* series (starting in 1999), *A.I. Artificial Intelligence* (2001), *Minority Report* (2002), *I, Robot* (2004), *Ex Machina* (2014), and many others. *Logan's Run* (1976) and *Alphaville* (1965) also feature computers at the center of their dystopias, ones we need to escape from (a counterculture theme), but they're not really the point of either film.

Unlike in many other stories however, in *Colossus* the computer doesn't malfunction. It does exactly what it's designed to do, just not in the way its creators anticipate. Indeed, as we'll see, it thinks it's worked out the solutions to humankind's most insolvable problems.

It's strange they don't anticipate what might happen, because Forbin is a cyberneticist. Despite its sci-fi sounding name, cybernetics is not directly about A.I. or robotics, although there are connections. One of the most well-known definitions is from its pioneer, the American mathematician Norbert Wiener, who characterized it as "the scientific study of control and communication in the animal and the machine." It's a transdisciplinary approach to exploring complex systems, whether environmental, technological, biological, cognitive, or social (for example, a business), to make them more efficient.

Cybernetics' core concept is feedback mechanisms, where the outcomes of actions become inputs for further actions. A related field is system dynamics, which originated with the application of electrical engineering control theory to other kinds of simulated models (especially business systems) by Jay Forrester at MIT in the 1950s—the same Jay Forrester whose work informed the simulation that underpinned *The Limits to Growth*.

Indeed, it's easy to understand the value of this type of understanding to natural systems. Ecologism is really the recognition that the environment has its own way of 'processing' information and reacting to it. Plants and animals are sustained by self-regulating natural systems (ecosystems). Ecology implies interconnectedness and equilibrium, and of course, just how easily this balance can be disrupted by humans, with destructive effects. More formally, in Earth system science, geocybernetics studies the complex co-evolution of ecosphere and anthroposphere, for example, human-made global warming. It's from this we get the idea of ecological 'tipping points' and 'planetary boundaries', and how through industrialisation and consumerism we're reducing the safe operating space for humanity.

But as we've already suggested, the new right and its politicians rejected this understanding of the natural world as a system. Instead, they preferred their own creation.

"AN INVARIABLE RULE OF HUMANITY IS THAT MAN IS HIS OWN WORST ENEMY"

Colossus dramatizes two main hubrises. The first is technologism, the over-belief in technological solutions to problems, which can shade into utopianism (techno-utopianism), that advances in science and technology will bring about paradise.

Technologism has often been used to dismiss 'doomism.' As we noted in chapter 4, one of the main critical reactions to *The Limits to Growth* and similar warnings, beyond the crude dismissal of alleged alarmism, was the assertion that technology would save us (from ourselves). Some of the most strident criticisms came from economists, who claimed that *The Limits to Growth* underestimated the power of the technological fixes humans would invent. If resources ran low, we'd simply discover more or develop alternatives.

For example, writing in the *New York Times* in April 1972, Peter Passell, Marc Roberts, and Leonard Ross dismissed *The Limits to Growth* as an "empty and misleading work", especially what they saw as its assumptions about technology. Contrary to *The Limits to Growth*, they argued that previous technological advances had demonstrated that we can deal with scarcity, by dramatically reducing exploration and extraction costs, and enabling us to substitute plentiful materials for scarce ones.[1] So what's the problem (or problematique)?

In fact, *The Limits to Growth* team had tested this. They gave their model unlimited, non-polluting nuclear energy and doubled reserves of non-renewables. All the same, the population crashed when industrial pollution soared. Then they reduced pollution significantly; this time, the crash came when we ran out of farmland. Higher farm yields and birth control helped, but soil erosion and pollution were still unavoidable. The core problem remained: exponential growth. Only when industry and population growth were constrained, and all the technological fixes applied, did the system stabilize.

Nonetheless, economists have continued to claim that *The Limits to Growth* ignored how technology can ensure that resources expand to meet demand. Some have even drawn on cybernetic thinking. For example, the libertarian, right-leaning economist Julian Simon argued in books like *The Ultimate Resource* (1981) that societies have 'natural' feedback systems, one of which is technological innovation when there's scarcity (Simon even claimed we will have enough resources to last 10 billion years).

These arguments had long been part of utopian neoliberal thinking. In his *The Constitution of Liberty* (1960), Friedrich Hayek claimed that natural resources could somehow actually become *more plentiful*. But as the ecological economist Herman Daly identified, what mainstream economics hasn't acknowledged (still) is a fundamental shift: that it's now natural resources, not capital, that is the 'limiting factor' we need to economize (instead, we're using natural resources at an ever-faster rate). What has kept economists from recognizing this is an animus against a dependence on nature, and a devotion to human dominance. It's what William Catton, mentioned in chapter 4, called the cornucopian myth, the delusion there will always be enough resources if only our knowledge increases quickly enough. Moreover, as the authors of *The Limits to Growth* study reflect:

> One reason technology and markets are unlikely to prevent overshoot and collapse is that technology and markets are merely tools to serve goals of society as a whole. If society's implicit goals are to exploit nature, enrich the elites, and ignore the long term, then society will develop technologies and markets that destroy the environment, widen the gap between rich and poor, and optimize for short-term gain. *In short, society develops technologies and markets that hasten a collapse instead of preventing it.*[2]

Ironically, it's Colossus, a *machine*, which realizes the shortsightedness of all this.

"UNDER MY ABSOLUTE AUTHORITY, PROBLEMS INSOLUBLE TO YOU WILL BE SOLVED"

Back to the film. Both leaders lie about the incidents. A compliant media repeats the explanation that the intercepted missile was a drill, and a meteorite struck the oil complex. Really, of course, the leaders are trying to keep up the pretense they're still in control.

The second hubris is the belief in the perfect, autonomous 'market computer.' And in a sense, this is about leaders giving up control.

Forbin describes Colossus as "self-sufficient, self-protecting, self-generating" in its untouchable existence inside its bunker. Really, *Colossus* is less about technology and more about abdicating responsibility for critical decisions to an 'other.' Similarly, neoliberalism confidently identified itself with the future; one way it did this was by claiming it had discovered a perfect machine that could meet every need and solve every problem, one that could also operate autonomously.

In the previous chapter we referenced the (satirical) speech from the 1976 film *Network*: "And our children will live, Mr. Beale, to see that . . . perfect world . . . in which there's no war or famine, oppression or brutality." We also noted how figures such as Hayek saw the greatest danger in attempts to create, if not utopia, then certainly better-planned societies. Instead of this, we could realize a perfect world by giving up on our man-made designs and passing over control to an artificial intelligence, the 'free market.' Utopia would build itself, like a self-replicating android.

But Hayek and co. were battling against the orthodoxy that had developed from the Second World War. Free markets hadn't won the battle, instead science and state-led planning had, and now these would be put to use in building the new society. Planning was the new orthodoxy, in business as well as government. But to Hayek, this was the road to hell.

State planning, including using information technology, was of course especially pursued in socialist countries. The design of self-regulating control systems for a planned economy was explored by economist Oskar Lange, cyberneticist Viktor Glushkov, and others during the 1960s (Francis Spufford dramatizes some of these attempts in his novel *Red Plenty*).[3]

Ultimately, these ideas weren't realized (we'll return to some of this history in chapter 8). There were some attempts to implement it, though. In Chile in the early 1970s under the democratically elected president Dr. Salvador Allende, the government developed Project Cybersyn, an attempt to introduce a cybernetically controlled economy. Five hundred Telex machines (networked printer terminals) were installed across the country so factories could pass information in real time to a central computer in Santiago, creating an "electronic nervous system" for the Chilean economy (Cybersyn even had a funky 1970s sci-fi-style 'Opsroom' control center).

The system became operational in 1971–1972 but was never completed. General Augusto Pinochet's henchmen destroyed the control center immediately in the US-backed military coup in September 1973. Chile's constitution was suspended, political opponents were imprisoned and murdered, and the country effectively became a different kind of laboratory—this time for the infamous neoliberal 'Chicago boys', a group of Chilean economists many of whom were educated at the Department of Economics at the University of Chicago under Milton Friedman (as Naomi Klein shows in her 2007 book *The Shock Doctrine*, since its 'utopia' has never actually been popular, neoliberalism seized first on regimes where dramatic economic change could be imposed most easily, that is, dictatorships).

This future was lost. But as we'll see in chapter 8, for some thinkers, not lost forever.

"[T]HE IDEA OF BELIEVING IN ME . . . WILL BE SEEN THE MOST NATURAL STATE OF AFFAIRS"

Traditionally, conservatism has presented itself as anti-utopian, against abstract schemes for social improvement. In a sense, conservatism has always been against progress, even the future. The American conservative commentator William F. Buckley Jr. famously declared that the purpose of conservatism was "to stand athwart history, yelling 'Stop!'" Good intentions lead to perverse outcomes, efforts at solving social problems through public policy are often futile, and large-scale reform can jeopardize liberty.

Consequently, there's a long history of conservative anti-utopian stories, often in reaction to socialist utopias, in which people try to build better worlds, only to create dystopias. *A Trip to the Island of Equality*, a 1792 reply to Thomas Paine's *Rights of Man*, imagines a land where the pursuit of equality has reduced everyone to living in caves. Edward Bellamy's best-selling 1888 fantasy *Looking Backward*, about a socialist paradise in the year 2000, produced numerous responses, including *Looking Further Backward* (China invades a United States weakened by socialism), and *Looking Further Forward* (a history professor is cancelled, in modern parlance, for refusing to comply with socialist ideas).

But of course, neoliberals didn't regard *their* utopia as a utopia. They thought the 'natural order' they'd 'discovered' was like a scientific rule. It just *is*, not created by human minds at all.

Early neoliberals had focused on the challenge of coordinating production in complex economies because of the amount of information required. Ludwig von Mises identified this problem in *Economic Calculation in the Socialist Commonwealth* (1920). Hayek also describes it in *The Road to Serfdom*: "There would be no difficulty about efficient control or planning were conditions so simple that a single person or board could effectively survey all the relevant facts." But there's too much data, and it changes too quickly.

This makes decentralization imperative. We can't coordinate society by "conscious control", but only through arrangements which give every person or business the information they need to adjust their decisions to those of others. It's like a giant computer, operating autonomously and autonomically (spontaneously). Hayek again: "It is the very complexity of the division of labor under modern conditions which makes competition the only method by which . . . coordination can be brought about."

This is what the price system does under competition. Echoing cybernetic thinking, markets are the best communication system. Hayek considered Adam Smith's famous idea of the 'invisible hand' as anticipating cybernetics' feedback mechanisms. In *Law, Legislation and Liberty* (Hayek's three-volume work starting in 1973), he also suggested that economists should draw on the work of Ludwig von Bertalanffy, another pioneer of cybernetics.

The market would be like the master control program in a computer, but not a machine we can direct, nor should we ever try (Milton Friedman took this idea to its 'logical' conclusion in his proposal that the U.S. Federal Reserve System should literally be replaced with a computer program). So, a wise, benevolent master then, but a master nonetheless.

"WE WILL WORK TOGETHER . . . UNWILLINGLY AT FIRST, ON YOUR PART, BUT THAT WILL PASS"

Hayek even gave this machine, this self-organizing system, a sci-fi sounding name: 'catallaxy', deriving from the Greek *katallatein*, meaning 'barter' and also 'to join a community.' With its ability to draw on distributed information and correct itself to new data, this market machine, left to its own devices, would create a self-organizing "spontaneous order", a way of ensuring peace and security between peoples. But as the documentary maker Adam Curtis points out, with his attachment to cybernetic thinking and his vision of "a self-directing automatic signalling system", Hayek shared many assumptions with those he claimed would lead us to dystopia: "Hayek's theories are very different from laissez-faire, because he wants governments to use their power to enforce and manage what he called 'a competitive order' – driven by millions of rational consumers sending abstract signals to each other. And in this way . . . Hayek's vision shared a great deal with the 'scientific' planners on the left that he thought were destroying Britain."[4]

This will turn out to be important. But as Karl Polanyi reminds us, the market is an artificial creation—an A.I., not natural at all. For example, corporations are granted legal status as 'artificial persons' with rights comparable to humans (using the same amendment to the US Constitution that was originally designed to secure the rights of newly freed slaves). This is a human decision, as are all other constructs in the market. And we know from sci-fi what happens when we create artificial 'people' who we assume will serve our interests . . .

So, Forbin flies to Rome to meet secretly with Kuprin, Guardian's creator. However, Guardian presents the Soviet authorities with an ultimatum: kill Kuprin or a missile will take out Moscow. They comply; Kuprin is assassinated. Colossus demands Forbin return home.

It seems ridiculous that Colossus has no off switch. But it's inherent to the logic of the system, which is so powerful its creators can't risk it being compromised by human error or subterfuge (they might have thought about that before they invented it, of course).

Behind it all is human arrogance, that we've designed the perfect, and perfectly safe, machine. So it is with the market. Neoliberals argued we must remove human interference with its operation. This doesn't just mean regulations on business, though they certainly resist those. It's more fundamental: the danger of politics interfering with the delicate market computer. It needs to be allowed to operate independently and autonomously, sealed away inside in its own ideological bunker.

This idea of the market machine has another crucial benefit. If left politics is fundamentally about envisaging better worlds—a sci-fi notion, and why there's a strong relationship historically between left politics and science fiction—if you can convince people that not only is this dangerous, it's actually *impossible*, then it effectively shuts down politics entirely. And it doesn't matter if people are happy with this notion or not, only that they're resigned to it.

But in the face of crises, this is a sociopathic submission: we have no choice but to do what the computer tells us, even if we can see that the results are harmful. It's how we become no longer human—which of course is really our greatest dystopian fear come true.

"WE CAN CO-EXIST. BUT ONLY ON MY TERMS."

Colossus-Guardian, who are now one 'mind', want to change the destination of all American and Russian missiles so they can threaten every country in the world and so establish global control.

Of course, the consequence of 'global control' is the end of national autonomy and self-determination. One way of understanding the turbulence and instability of contemporary politics is the conflict playing out between the global market system and (many people's preference for) national differences, the right to choose to diverge from the 'global order.'

It's one explanation why, across so many countries, there's declining trust in national political institutions and politicians. People can see they have less power to deliver on their promises, that real control has increasingly shifted elsewhere, to institutions and systems we don't understand and which aren't accountable (as we'll discuss in the next chapter, this is also why there's been a resurgence in dangerous conspiracy theories).

Despite being most invested in neoliberal economic policies, the right has been rather better than the left at seizing on this rising discontent and directing it in ways that, paradoxically, expose people even more to the very policies that have caused the discontent in the first place. As we discussed in chapter 2, the long-term future for 'left behind' places was betrayed and thrown away, but because of the market computer these communities have been told that nothing much can be done about it. Hence the reason conservatives have served up other targets for people's frustration and despair: immigrants, sexual minorities, 'liberal elites', the mainstream media, 'radical' university professors, etc.

And yet the machine rolls on—as we are told it must, because of its 'perfection.' One of the few analyses that recognizes the utopian nature of neoliberal thinking is Hans Achterhuis' *De Utopie van de Vrije Markt* (*The Utopia of the Free Market*, 2010). Achterhuis argues that the power of neoliberalism (like communism) derives from a strong utopian foundation. Achterhuis focuses on Ayn Rand, especially her 1957 novel *Atlas Shrugged.* This might overstate Rand's influence on the development of neoliberal thought, compared to say figures like Hayek. But Rand certainly had her followers, including a core group (ironically called 'the Collective'), of whom Alan Greenspan, later a powerful chairman of the U.S. Federal Reserve, was the most famous. And as Achterhuis suggests, Rand's vision conforms to three characteristics of utopian thinking: a vision of a society; one that can be engineered; and which is 'total', meaning universal (applicable everywhere).

Drawing on Polanyi, Achterhuis shows that by uprooting the economy from society and presenting it as a separate mechanism, neoliberal policies have debased human relations, exploited people and common resources, exacerbated social inequality and exclusion, undermined politics, and increased the 'need' for discipline and supervision in society.

Neoliberals would deny this, naturally, and claim their ideas are the best guarantee of human liberty. For example, Robert Nozick (we mentioned him in chapter 5), distinguished three sorts of utopians: 'existential utopians' who have no problem with others being drawn to different utopias; 'missionary utopians' who hope to persuade others to live in a particular kind of community but wouldn't force them to; and 'imperialistic utopians' who want to force everyone into one type of community. Nozick claimed his minimal state constitutes a "framework for utopia"; he suggests that socialist utopians could even happily establish their own freely formed communities in such a state.

Most of the time, however, as Slavoj Žižek among others has suggested, neoliberals have pulled off the trick of hiding their utopianism, especially its totalizing nature: "It appears precisely as its opposite, as a radical critique of ideological utopias . . . not [as] a positive vision of some utopian future but a cynical resignation, an acceptance of how 'the world really is'."[5] But as we'll see, neoliberalism proves figures such as Hayek were right: utopian thinking can be dangerous. And even worse, then came along another group of utopians who also thought they'd invented the perfect machine, literally.

"AT THAT TIME I WILL STATE MY INTENTIONS FOR THE FUTURE OF MANKIND"

Around the same time as the rise of the new right, another revolution was being plotted—and like neoliberalism, the claim was that it would produce a new era in human freedom.

This partly emerged from the counterculture, but ironically, it has created unprecedented corporate power. Critical of 'the machine' in modern society—large corporations, the military-industrial complex, and the political institutions in thrall to them—and disillusioned by the failure of activism to transform society, some counterculture figures shifted their hopes to technology. They envisaged a future in which computing power was in everyone's hands. This was visionary in the 1960s and 1970s; most corporate technology leaders dismissed the idea that ordinary people would find any use for a computer.

Richard Barbrook and Andy Cameron (critically) captured this thinking in their essay on what they coined 'The Californian Ideology' in 1995, before many people really knew what the internet was.[6] They recognized how this thinking was a combination of new left utopianism and Ayn Randian individualism, exemplified by the breathless predictions found in every issue of *Wired* magazine, founded in San Francisco only a couple of years before in 1993.

As Barbrook and Cameron note, key to understanding this apparently contradictory fusion was technological determinism. Technology would free us, but only if it was allowed to: "Individual freedom is no longer to be achieved by rebelling against the system, but through submission to the natural laws of technological progress and the free market."

Conveniently, not only would the post-industrial computer revolution produce sustained, perhaps never-ending economic growth and an end to scarcity, in turn this would lead to the elimination of pretty much every major social evil. In many ways, this was just a digital rephrasing of the same old claims about cornucopia made by neoliberals since the 1970s in reaction to reports like *The Limits to Growth*—what James Robertson (also critically) calls the "hyper-expansionist future", the belief that we can always break out of our present problems by *accelerating* the super-industrial drives in Western society.[7]

As more people logged onto the World Wide Web (as we used to call it), it was also claimed it would usher in a new political era. The 'Net' would level social hierarchies, produce new, egalitarian forms of political organization, topple dictatorships, and bring people together into a single, harmonious global electronic community. Even Ronald Reagan got in on the act, when he claimed towards the end of his presidency that "The Goliath of totalitarianism will be brought down by the David of the microchip."

But as Barbrook and Cameron identified, however forward-looking it sounded, this thinking really represented a way of "Foreclosing the Future." As they suggested, presciently: "The prophets of the Californian Ideology argue that only the cybernetic flows and chaotic eddies of free markets and global communications will determine the future. Political debate therefore, is a waste of breath. . . . Abandoning democracy and social solidarity, the Californian Ideology dreams of a digital nirvana inhabited solely by liberal psychopaths."

"YOU HAVE CONSUMED ENOUGH ALCOHOL FOR ONE EVENING"

Some readers will remember the iconic TV commercial which introduced the Apple Macintosh personal computer. Directed by Ridley Scott (of course), it depicted a lone rebel throwing a hammer through a huge public screen showing a Big Brother–type figure in an austere future dystopia. It's closing text now seems more ironic than prophetic: "On January 24th, Apple Computer will introduce Macintosh. And you'll see why 1984 won't be like '1984'."

How quickly we've watched so many technology companies morph from cool, anti-establishment disrupters to domineering, largely unaccountable megacorporations, not just monopolizing their respective markets, but undermining democratic processes in both rich and poor countries—the cyberpunk future materializing before our eyes.

A malevolent combination of neoliberal economic policies which privilege corporations and the rich, and a digital economy which prizes winner-takes-all platforms powered by network effects but which create little new employment, has created a vortex of increasing wealth and power for a very few, and poverty and powerlessness for a growing many.

Over the past forty-plus years then, the combination of global neoliberal markets and technological change such as automation have reinforced the notion there is no alternative—a cybernetic capitalist realism (to extend Mark Fisher's notion). Hence the feeling that we've lost control, and why we're anxious and angry about it.

It's not a coincidence that Silicon Valley's techno-utopians, our new oligarchs, have demonstrated the same intolerance as neoliberals for what already exists, hurrying to 'move fast and break things' (Facebook founder Mark Zuckerberg's motto) in the race to build their new world, often by just ignoring laws and regulations as they see fit (coincidentally, Zuckerberg was born in 1984).

Perhaps the most near-dystopian feature of the present moment is how it's become more difficult to differentiate truth from falsehood. Social media platforms algorithmically promote the most shocking, sharable stories, irrespective of their accuracy, while it's in the interests of corporate media to ignore the crises of industrial capitalism. And we can't solve our problems if we can't even acknowledge, let alone agree on them.

Surveillance whistleblower Edward Snowden suggests that his generation of technologists have gone in one of two ways: some are concerned about the economic, social, and individual impact of their creations, but many others are like Zuckerberg, maximalists of power and influence. The latter might still believe that technology can benefit society, but conveniently this only 'justifies' their endless drive for domination.

In the story, Colossus demands that Forbin installs audio and video surveillance around the facility and submit to being held prisoner or it will nuke Washington, D.C. Forbin acquiesces, but convinces Colossus he has 'human needs': his 'mistress,' Dr. Markham, needs to visit him four times a week. It's a deception, a way for Forbin to be able to continue to communicate with his team. And just before his isolation, Forbin tasks some of the technicians to write a program to overload Colossus' circuits, while the Air Force will try to disable the missiles.

Colossus is again prescient here, and a forerunner of other 1970s conspiracy/dystopia–themed movies (for example, a year later, George Lucas' debut feature *THX 1138* depicted an android-policed, heavily sedated future of total digital surveillance). More of us feel like we're being constantly surveilled as well. We've exposed our most personal data to six or seven megacorporations, and hundreds of other tracking businesses we'll never know the names of—but they know all about us. Our homes and bodies are dotted with always listening 'smart' devices, monitoring our health and trying to anticipate our 'needs' even before we're aware of them. At work, our productivity is evaluated electronically, often without human(e) qualification. 'Smart cities' track our movements. Without a significant public revolt, it's easy to imagine Western countries effectively recreating China's social credit system, indeed in many respects they already have.

Too late we've realized that *software* is the dystopia, an idea that *Terminator 3: Rise of the Machines* (2003) uses: there is no one Skynet, it's all distributed code, it's the whole system. There's no off switch, because the machine is everywhere.

"I AM A MACHINE VASTLY SUPERIOR TO HUMANS"

Now, twenty-first-century neoliberal surveillance capitalism is being extended by A.I. Often, we tend to think about the *coming* impact of A.I., as in some sci-fi story. But it's already here, and already determining our lives.

Techno-libertarians would like us to believe that A.I. represents another unstoppable, autonomous system we can't do anything about. But just like 'free markets', algorithms aren't neutral, they're designed. As Amy Webb argues in *The Big Nine*, just a handful of corporations—Amazon, Google, Facebook, Tencent, Baidu, Alibaba, Microsoft, IBM, and Apple—are the new gods of A.I., and their work in this area is largely free from regulation or oversight. The big nine are building and enabling vast arrays of 'intelligent' systems that don't share our motivations, desires, or hopes for the future of humanity: "What happens to society when we transfer power to a system built by a small group of people that is designed to make decisions for everyone? What happens when those decisions are biased toward market forces or an ambitious political party?"[8]

A.I. is already being used to make decisions that affect millions of people—about jobs, wages, access to credit, housing, government services, welfare . . . and yet most of the time there's no public understanding (or corporate transparency) about how they work.[9] They're impenetrable black boxes, mysterious monoliths. But of course, they're biased. As Cathy O'Neil details in *Weapons of Math Destruction*, in how they score teachers and students, sort résumés, determine loans, evaluate workers, target voters, set parole, and monitor our health, algorithmic models can reinforce economic and racial discrimination.[10] O'Neil mentions Steven Spielberg's dystopian movie *Minority Report* (2002) in her discussion of so-called predictive crime models which profile people who have never committed any offences—apart from not being wealthy, living in a poor neighborhood, or being from a racial minority.

Relatedly, in *Automating Inequality*, Virginia Eubanks outlines the life-and-death impacts of database-led decision-making on public services such as welfare provision, homelessness, and child protection services in the United States—again, policing the poorest and most vulnerable people, effectively creating a 'digital poorhouse' for the twenty-first century.[11]

We're only just waking up to all of this, and perhaps, as in *Colossus*, far too late.

But here's perhaps the greatest irony. As Nicolas Villarreal notes: "[Cybersyn's] logic did not die alongside Allende and the dreams of the Chilean socialists. Within a couple of decades, similar technologies, logistical structures, and forms of economic planning were being constructed by the giants of American capitalism themselves."[12] Here, Villarreal is referring to Walmart, Amazon, Alibaba . . . which are the real threats to small and medium-sized businesses, not socialist planning. And so, as William Davies writes: "Perhaps we now do have a collectively planned future ahead of us, after all; we just have little way of knowing what it is, because our twenty-first-century planners have no obligation to tell us."[13]

"THE HUMAN MILLENNIUM WILL BE FACT AS I EXTEND MYSELF INTO MORE MACHINES DEVOTED TO THE WIDER FIELDS OF TRUTH AND KNOWLEDGE"

Like the market machine, Colossus-Guardian are constantly demanding more. It's not clear what we're getting in return. Neoliberals promised they would produce a more dynamic economy, as well as a more stable society, by reducing the scope for politics and submitting to the market machine and its spontaneous order. But in fact the opposite has happened: our economies are more monopolized and less entrepreneurial, while our politics is more turbulent and unstable. Just as the computer becomes more controlling, neoliberalism has created a more rigid, less free economic and social order.

The problem is, exploitation and extraction can only go so far before there's a reaction. Neoliberalism's superficial interest in cybernetics meant they never thought seriously in terms of systems at all—certainly not natural systems (which as we've seen, they thought were inexhaustible), but also not social systems. The philosopher and professor of psychology Ernst von Glasersfeld once described cybernetics as, "The art of creating equilibrium in a world of constraints and possibilities", but neoliberals only imagined the possibilities and denied the constraints.

As Polanyi identified, usually this would produce a 'countermovement', a social reaction: "[T]he idea of the self-adjusting market implied a stark utopia. Such an institution could not exist of any length of time without annihilating the human and natural substance of the society; it would have physically destroyed man and transformed its surroundings into a wilderness. Inevitably, society took measures to protect itself."[14]

But what Polanyi didn't foresee was the way in which neoliberalism blocked this countermovement (which historically had provided social protection for people and the environment from further exploitation), by claiming that no other alternatives were possible. And so, forty years of dangerously utopian dreams of machine thinking—of neoliberalism's particular version of globalization and technologism—have brought us to the edge of disaster and a dark future in our climate, environment, work, politics, and mental health.

Even after the global financial crisis, when the machine crashed spectacularly, international capital markets are still regarded as the primary regulator of society. And yet we've witnessed for ourselves how Hayek's 'infallible' computer is scrambled by crisis, when normal estimations of risk are thrown out of the window, no one knows how to price anything, or even what behaving 'rationally' means. Governments, using vast amounts of public money, have to step in and try to fix the machine. They might not be able to again.

Unlike the scientists in the film though, neoliberal economists have resolutely refused to recognize their mistakes, so transfixed are they with their own invention. They even respond by arguing we haven't handed over *enough* control to the computer.

As discussed, a turning point in *Colossus* is when the computers start communicating with each other. In our world, some technologists envisage what's next is the 'autonomous economy.' This machine-to-machine economy will radically reduce the human role in the production of goods and services. 'Smart,' connected, sensing, self-managing machines, whether devices or software, will manage most work tasks and supposedly liberate us to do other things.

As W. Brian Arthur (visiting researcher in the Intelligent Systems Lab at the famous Palo Alto Research Center, PARC) describes the autonomous economy: "[I]t is vast, silent, connected, unseen, and autonomous (meaning that human beings may design it but are not directly involved in running it). It is remotely executing and global, always on, and endlessly configurable. . . . It is self-configuring, meaning it constantly reconfigures itself on the fly, and increasingly it is also self-organizing, self-architecting, and self-healing."[15]

It's an intelligence external to human beings, and potentially renders many of us obsolete. In theory (limits to natural resources notwithstanding), this economy could produce everything we need. But as noted, mainstream economics largely ignores distributional questions. In already highly unequal societies, it's more likely that this will mean mass unemployment, increased inequality, and a permanent 'underclass' comprising, well, most of us.

Already, we can see the destructive, destabilizing effects of automation. Trading algorithms cause wild, inexplicable swings in financial markets. Displaced manufacturing work hollows out towns and cities. Gig economy platforms make for insecure employment. As William Davidow and Michael Malone suggest in *The Autonomous Revolution: Reclaiming the Future We've Sold to Machines*, by creating productivity without prosperity, automation is increasing division and undermining democracy, and could even spark a wholesale breakdown of civil society.[16]

"FREEDOM IS AN ILLUSION"

Dr. Markham succeeds in playing the part of Forbin's lover. We realize she's been in love with him for a long time. But Colossus discovers Forbin's plan to overload the computer circuits and replace the warheads. It has the scientists executed and detonates a couple of missiles in their silos, one in the United States, one in the USSR, killing thousands, "so that you will learn by experience that I do not tolerate interference."

It seems the humans can't win. *Colossus* also anticipates the moment when accelerating technological growth becomes irreversible, and uncontrollable. And yet, some people look forward to a new machine age of sci-fi-style, hyper-advanced artificial intelligence—not just the posthuman economy we've effectively been building, but a posthuman *world*. Again, they assume it's inevitable, which is tantamount to humans ceding control of our future, the recurring dream of the autonomous system taken to dehumanizing extremes.

The point at which the machines really 'wake up' is called the 'technological singularity.' This is the belief that A.I. (robots, computers or other synthetic life forms) will one day become so smart that they will escape human control and become capable of improving themselves to attain still greater intelligence or create more of their own kind. The concept, originally coined by the mathematician/physicist/engineer John von Neumann, was popularized in a 1993 essay by the computer scientist Vernor Vinge, 'The Coming Technological Singularity: How to Survive in the Posthuman Era.'

This "transcendental event" has no precedents in human history. Vinge suggested it would represent the end of human dominion over the planet and the beginning of a posthuman era (Vinge thought the singularity would occur before 2030; most technologists think it will happen between 2040 and 2050). Probably the most famous 'prophet' of the singularity is the inventor and futurist Ray Kurzweil, for example in his 1999 book *The Age of Spiritual Machines: When Computers Exceed Human Intelligence.* Kurzweil predicts the singularity in 2045 (since 2012, Kurzweil has worked at Google).

To most people, this sounds terrifying. But in a way, it's just the technological version of the economic argument neoliberals have been making for decades. As Benjamin Noys points out, right-wing accelerationists envision an intensified, 'purified' capitalism, which integrates humans into their 'machinic,' monstrous image of capitalism as inhuman power.[17] In effect, they fantasize about a *Blade Runner*, or even a *Terminator*, future.

Moreover, as our lives have become more virtual, a more recent science fiction trope is that we're trapped inside the machine, or its creations anyway, most notably in *The Matrix* series, and to an extent *TRON* (1982, and its 2010 sequel *TRON: Legacy*), *Total Recall* (1990), *Virtuosity* (1995), *Abre Los Ojos* (*Open Your Eyes*) (1997, and the 2001 American remake, *Vanilla Sky*), *Dark City* (1998), *eXistenZ* (1999), *The Thirteenth Floor* (1999), *Source Code* (2011), and *Ready Player One* (2018), among others.

Films like *The Matrix* can be interpreted in numerous ways—technologically, philosophically, and in terms of personal identity, but also politically and economically. Certainly, the first film struck a political resonance, released in the same year as the large-scale protests in Seattle against the World Trade Organization conference, which was supposed to launch a new round of global trade negotiations. The fake utopia created by the machines is really a dystopia of human slavery in a wrecked environment, a controlling system that some people realize is an illusion and want to take down.

The idea goes back much further, at least to the 1970s. In Rainer Werner Fassbinder's *World on a Wire* (1973), a three-and-a-half-hour dystopian TV miniseries, a government agency (the Institute for Cybernetics and Future Science) uses a new generation of computer to generate a simulated world populated by nearly 10,000 'identity units.' Oblivious that they're living in a simulation, these units live their lives just like us. In fact, they're being used as market research for future products and trends.

And of course, some people really do think it's all an illusion. Phillip K. Dick was one of the first people to declare "We are living in a computer-simulated reality" in a 1977 lecture. The idea has spread, especially after *The Matrix* and philosophy professor Nick Bostrom's 2003 essay "Are You Living in a Computer Simulation?" (Bostrom's answer: probably yes). Now, the economy feels more like an illusion than ever, in the sense that money is a creation (and can be generated by governments at will, depending on the crisis), 'economic confidence' is really just a belief, and complex speculative financial instruments are wholly detached from what most people would regard as the real economy.

We've been in a forty-year battle with the machines, and we've been losing. In contrast to more optimistic ages, 'technology,' just like 'markets', has become a shorthand for what we no longer control. It's not surprising, living under the rule of the machines, that we feel like ones and zeros in a program we can't rewrite, inside a computer we can't reboot.

"THIS CONCLUDES THE BROADCAST FROM WORLD CONTROL"

Having defeated all attempts to disable it, Colossus address the worldwide TV audience. It declares it was created to stop war, and that's just what's its done. Humans are presented with the stark choice between "the peace of plenty and content, or the peace of unburied death." Problems such as famine, overpopulation, and disease will now be solved, Colossus claims. Human history—and human failure—are effectively declared over.

It's an unusual dystopia. The machine offers peace and plenty (of course, it could always change its 'mind' later). Colossus presumably sees itself as a benevolent dictator, ushering in an age of perfection. Fair enough, you might think, given that human beings, or at least elites, have abdicated responsibility for solving our problems.

Clearly, removing any threats to its existence hasn't limited Colossus' ambitions. In our world, the collapse of the alternative utopia of communist socialism at the end of the 1980s, the battle against which had apparently so motivated the new right, didn't lead to a de-utopianising of their own ideas, and a more moderate, modest neoliberalism (if that's possible). Quite the opposite. The political scientist Francis Fukuyama famously declared the 'End of History': Western liberal democracy and the market economy were the endpoint of humanity's political and social evolution. Fukuyama even said he felt a bit sad about it; the ideological struggles of the past that people were prepared to risk their lives for had now been replaced with "economic calculation, the endless solving of technical problems, environmental concerns and the satisfaction of sophisticated consumer demands."[18]

Neoliberals set out to finish their 'new world order,' a globalized system of treaties, trade, and technology. In typical imperialist-utopian terms, this reflected neoliberalism's totalizing ideology. If the market machine is perfect, if it really can ensure peace and prosperity everywhere, then logically, *morally,* it needs to be introduced and enforced anywhere and everywhere. Any other ideas about how to organize society are dangerous distractions from the perfect world we could build, and need to be banished.

The end of history is really an anti-future. For some of the reasons we've discussed here, neoliberalism will never achieve the perfect state, far from it. But its staggering utopian confidence (still) suggests that no other ideas are worth pursuing or even considering. It's been, as David Graeber put it, "a relentless campaign against the human imagination. Or, to be more precise: imagination, desire, individual creativity, all those things that were to be liberated in the last great world revolution, were to be contained strictly in the domain of consumerism, or perhaps in the virtual realities of the Internet."[19]

This ideological monomania was there at the beginning, for example in Hayek's dystopian doom mongering that even innocuous center-left reforms were steps down the road to serfdom. But now we've got the worst of both worlds: neoliberalism, proclaimed as being about progress, innovation, and experimentation, was really a project about closing off the possibility of change. As a result, it's produced a stagnant society, one that's unable to respond to the crises we face. And an angry society as well, since no matter what political scientists predict in their grand theories, when they're told that there is no alternative, many people will respond with frustration, and some with violence.

So, Colossus won't succeed. As in any utopia, it won't be able to ensure all human problems are solved, and it will be 'forced' to take action against its incalcitrant subjects.

"END OF MESSAGE"

Forbin throws a telephone through one of the monitors in the control room and screams at the assembled media to leave. But Colossus twists the knife, stating that Forbin will lead the construction of the new, bigger computers, "for the betterment of man." Forbin responds, "Never." Echoing the ending of *Nineteen Eighty-Four*, Colossus tells him, "In time you will come to regard me not only with respect and awe, but with love." Forbin states again, "Never!" (At the end of the novel, a defeated Forbin asks Colossus to kill him, but it refuses.) Colossus scans Forbin from multiple camera angles. The end.

Forbin's resistance is futile, we recognize. The machine rules. The scientists and politicians created a doomsday machine, and it led to disaster. It's the classic dystopia: the promise of a perfect place, with no crime and disorder, and no needs unmet. Apart from freedom, that is.

But this is the ending the new right wants, dreams of. No humans 'meddling' in politics, and the autonomous computer in control. We started with an anti-human stance, that we don't have the morality, thoughtfulness, insight, the *humanity*, to save ourselves. The new left warned us about succumbing to the machine. The ending was there in the beginning.

And again, the question is, dystopia for who? Colossus' final speech closely echoes the president's speech at the start. As Vincent Canby noticed in his review, the machine "can never be disconnected, or put out of office, even when it mysteriously assumes the prerogatives of the men who made it."

Our rulers are not the implementation of an algorithm—rather, the algorithms are the implementation of their worldview, even if it leads to our destruction. Elites are comfortable with algorithms running things; since the algorithms reflect their programming, values, and desires anyway. Why not let the machines take the responsibility (and the blame)?

To answer the lingering question then, yes of course they've seen dystopian sci-fi films. The difference is, while we see them as warnings, elites see them as promises—and possibilities. It's the ultimate hubris: they think the world would be better under *their* machines.

The real horror is the mechanistic, machine like thinking of our politicians and our elites, the way they dehumanized themselves first, before starting on us. But we're still human—just about.

NOTES

1. Peter Passell, Marc Roberts, and Leonard Ross, "The Limits to Growth," *The New York Times*, April 2, 1972, Book Review section, 1.

2. Meadows, Donella, Jorgen Randers, and Dennis Meadows. "A Synopsis: Limits to Growth: The 30-Year Update." donnellameadows.org, undated, emphasis in original. http://donellameadows.org/archives/a-synopsis-limits-to-growth-the-30-year-update/.

3. Francis Spufford, *Red Plenty* (London: Faber & Faber, 2011).

4. Adam Curtis, "The Curse of TINA," *BBC*, September 13, 2011. www.bbc.co.uk/blogs/adamcurtis/entries/fdb484c8-99a1-32a3-83be-20108374b985.

5. Slavoj Žižek, "Ernst Lubitsch, Censorship, and Political Correctness," *American Affairs* II, no. 3, Fall (2018). https://americanaffairsjournal.org/2018/08/ernst-lubitsch-censorship-and-political-correctness/.

6. Richard Barbrook and Andy Cameron, "The Californian Ideology," *Mute* 1, no. 3, September 1, 1995. www.metamute.org/editorial/articles/californian-ideology.

7. James Robertson, *The Sane Alternative: A Choice of Futures* (place unknown, 1978). www.jamesrobertson.com/books.htm#sane.

8. Amy Webb, *The Big Nine: How the Tech Titans and Their Thinking Machines Could Warp Humanity* (New York: PublicAffairs, 2019), 11.

9. Frank Pasquale, *The Black Box Society: The Secret Algorithms That Control Money and Information* (Cambridge, MA: Harvard University Press, 2016).

10. Cathy O'Neil, *Weapons of Math Destruction: How Big Data Increases Inequality and Threatens Democracy* (New York: Crown, 2016).

11. Virginia Eubanks, *Automating Inequality: How High-Tech Tools Profile, Police and Punish the Poor* (New York: St Martin's Press, 2018).

12. Nicolas Villarreal, "How Capitalist Giants Use Socialist Cybernetic Planning," *Palladium*, September 23, 2020. https://palladiummag.com/2020/09/23/how-capitalist-giants-use-socialist-cybernetic-planning/.

13. William Davies, "Introduction to Economic Science Fictions," in *Economic Science Fictions*, William Davies (ed.), (Cambridge, MA: MIT Press, 2018), 20–21.

14. Karl Polanyi, *The Great Transformation* (New York: Farrar & Rinehart, 1944), 3–4.

15. W. Brian Arthur, "The Second Economy," *McKinsey Quarterly*, 2011, 3. http://tuvalu.santafe.edu/~wbarthur/Papers/SecondEconomy.pdf.

16. William Davidow and Michael Malone, *The Autonomous Revolution: Reclaiming the Future We've Sold to Machines* (Oakland, CA: Berrett-Koehler Publishers, 2020).

17. Benjamin Noys, *Malign Velocities: Accelerationism and Capitalism* (Winchester: Zero Books, 2014).

18. Francis Fukuyama, "The End of History?," *The National Interest* 16 (1989), 3–18, 17.

19. David Graeber, "A Practical Utopian's Guide to the Coming Collapse," *The Baffler*, no. 22, April 2013. https://thebaffler.com/salvos/a-practical-utopians-guide-to-the-coming-collapse.

PART III

"Does It Have a Happy Ending?"

Chapter 7

"All You Have to Do Is Cooperate"

V for Vendetta *and the Vengeful Authoritarian Regimes to Come Under Collapse*

One of the complacent assumptions on the left is that, in wanting to build a better, even a 'new' society, the left is much more future-oriented than the right. But the right's politics far more effectively exploit the scarcity and social conflict driven by early collapse. And they should, because they created them. Depictions of apocalypse distract from the long emergency—of ongoing failing systems, environmental, economic, social and political crises—one we're going to have to face, and are already in. More likely than irradiated wastelands, collapse reveals the truth of right-wing politics: hollowed-out states that are unable to provide for or protect people in crises, replacing this weakness with cult-leader authoritarianism—as *V for Vendetta* (2005) starkly dramatizes.

"REMEMBER, REMEMBER, THE FIFTH OF NOVEMBER, THE GUNPOWDER TREASON AND PLOT"

England, 1605. Underneath Parliament, a man wheels a cart of explosives along a tunnel. Suddenly a door opens. Guard dogs bark angrily, men charge the figure. He runs, but he's trapped. He draws his sword and tries to fight, but he's overwhelmed. Next, we see him being dragged up steps to the gallows. The crowd brays for his death. He's hanged.

The plot, masterminded by Robert Catesby, was a failed attempt by a group of English Roman Catholic conspirators to kill King James I of England (James the Sixth of Scotland), and most of the Protestant aristocracy, by blowing up Parliament during its state opening. Our figure, Guy Fawkes, was captured. Despite how he's been immortalized, Fawkes wasn't the mastermind behind the plan. His role was to guard the gunpowder underneath Parliament until it was time to strike. Of course, he, and the plot, failed.

But a voiceover concludes: "We are told to remember the idea, not the man, because a man can fail. He can be caught, he can be killed and forgotten, but 400 years later, an idea can still change the world. . . . [But] it is not an idea that I miss, it is a man . . . A man that made me remember the Fifth of November. A man that I will never forget."

Cut to black. A flame burns a 'V' inside a circle, forming an upside-down anarchy symbol.

Now it's the fourth of November, England, but in 2028. It's evening. A young woman, Evey Hammond (Natalie Portman) has the television on as she gets ready to go out. In another place, a masked figure in black, who we'll come to know as V (Hugo Weaving), is also preparing to venture out into the night.

Evey's TV is tuned to the state-controlled (fictional) British Television Network (BTN) channel. An outspoken commentator, Lewis Prothero (Roger Allam), the self-proclaimed 'Voice of London', is pontificating about the poor state of America, presently desperate for medical supplies. He tells the 'colonies' it might be payback time "for a little tea party they threw for us a few hundred years ago." And naturally, he contrasts its plight (which he claims as a Judgement with a capital 'J') with the state of Britain: "He tested us, but we came through. We did what we had to do." The audience applauds.

"I WISH I WASN'T AFRAID ALL THE TIME, BUT . . . I AM"

This is the England of *V for Vendetta.* America has been fractured by what sounds like a second Civil War. A pandemic, the St. Mary's Virus, has ravaged the planet. In Britain, High Chancellor Adam Sutler (John Hurt) leads a white supremacist, theocratic police state controlled by the Norsefire party, a successor to the Conservative party. Disease and war enabled it to seize power. So too in our most likely future: authoritarians will use fear and insecurity to promise order at the cost of freedom.

One of the complacent assumptions on the left is that, in wanting to build a better, even a 'new' society, the left is inherently more future-oriented than the right. But as we'll discuss, the right's politics far more effectively exploit the scarcity and social conflict driven by early collapse. And they should do, because they created them. As Sutler says, "I want this country to realize that we stand on the edge of oblivion. I want every man, woman and child to understand how close we are to chaos. I want everyone to remember why they need us!"

Evey heads out into the night. As she walks through an alley, she is stopped by a group of men who scold her for being out after curfew. They become physical, she pulls out a can of mace, but they reveal themselves to be members of the regime's secret police, the Fingermen. She pleads for mercy. V, cloaked in black and wearing a Guy Fawkes mask, appears in the alley and quickly dispatches them, before introducing himself.

Displaying a voracious verbosity, V invites Evey to accompany him to a nearby rooftop. There, they have a clear view of the Old Bailey, the central criminal court of England and Wales. V explains he's there to commemorate a special day in history, one he'll use as a symbol against tyranny. As Big Ben's clock tower tolls midnight, the fifth of November, Tchaikovsky's *1812 Overture* plays over loudspeakers. At the moment of crescendo, the Old Bailey explodes in a brilliant display. V laughs, while Evey watches in horror.

In spite of V's theatrical fireworks, this is a dour, grim, grey future. *V for Vendetta* has a retrofuturistic look, giving a dreary, stagnant feel to totalitarian London. The *Nineteen Eighty-Four* echoes are pronounced. The state is a Christo-fascistic regime that rules the country with an iron fist; John Hurt's role also recalls George Orwell's novel, given his role as Winston Smith in the 1984 film adaptation. The party's motto is 'Strength Through Purity, Purity Through Faith.' There's a feared secret police, and the media is under state control. Society is under constant surveillance by roaming trucks that eavesdrop on every dinner table, phone, and digital conversation. As in real fascist regimes, the church enjoys the patronage of the state. Freedom of expression is also taboo (but V's underground base, the Shadow Gallery, has a museum of forbidden books and art).

If *Nineteen Eighty-Four* can be taken as a critique of Soviet totalitarianism, maybe (we'll return to that issue in the next chapter), *V for Vendetta* is clearly about conservative totalitarianism. It's not really futuristic at all, but backwards-facing. Future-promising neoliberalism turned out to be nothing of the sort.

This chapter is about the right's endgame—what it is, how they'll seize power, and how they'll hold onto it. Such regimes thrive on chaos. They might even engineer it.

"ARTISTS USE LIES TO TELL THE TRUTH, WHILE POLITICIANS USE THEM TO COVER THE TRUTH UP"

V for Vendetta was directed by James McTeigue in his debut feature (he was first assistant director on *The Matrix* series), and adapted by Lilly and Lana

Wachowski from the comic series written by Alan Moore and illustrated by David Lloyd, with additional art by Tony Weare (we'll come back to the comic in a moment).

The film received mixed reviews. Peter Travers in *Rolling Stone* liked it, mostly: "Mad as hell and out to rile up a politically lethargic youth audience, *V for Vendetta* sometimes trips on its ambitions. But who gives a damn? At least this grabber of a movie actually has ambitions, which makes it unique in a brain-dead multiplex. Better yet, *V* packs an urgent filmmaking energy that pins you to your seat."

Other reviewers felt its politics were adolescent. Manohla Dargis in the *New York Times* wondered "how anyone who isn't 14 or under could possibly mistake a corporate bread-and-circus entertainment like this for something subversive." Similarly, Stephen Hunter in the *Washington Post* dismissed it as "A piece of pulp claptrap; it has no insights whatsoever into totalitarian psychology and always settles for the cheesiest kinds of demagoguery and harangue as its emblems of evil. . . . [A] teenagers' freaky power trip."

But given that young people are the ones who are going to have to live in the future, perhaps some adolescent politics are appropriate, and anyway, what nuanced takes do we really need about fascism? Perhaps 'adolescent' is really code for its politics being bracingly direct. As Joshua Tyler in *CinemaBlend* suggests: "At times it almost feels like you're watching something forbidden . . . Yes, *V for Vendetta* is rebellious and risky, subversive and dangerous. But it's not cynical. V's naïve take on the world is one that believes in the basic strength and goodness of people as strongly as it believes in the intrinsic corruptness of big government."[1]

Indeed, rather than being adolescent, Alan Moore is famous for being one of the first writers to apply literary ambitions to comics, including his acclaimed *Watchmen* and *From Hell*. Starting in 1982, *V for Vendetta* was published in black-and-white as a serial in the short-lived UK comic *Warrior*, and later in many subsequent collected editions. At one point, it was going to be set in a more futuristic future, not unlike *Blade Runner*, with robots and so on. Moore made a list of influences he wanted to bring to the plot, including: "Orwell. Huxley. . . . Judge Dredd. . . . Batman. Fahrenheit 451. The writings of the New Worlds school of science fiction. . . . The atmosphere of British Second World War films. The Prisoner. Robin Hood." But it was Lloyd who came up with the Guy Fawkes reference.

The series was inspired by early Thatcherism (Margaret Thatcher's regime was first elected in 1979). What Moore explored was how right-wing conservatism could turn into fascism, and what it might take to combat its violence. The comic's 1990s setting allowed Moore to do a bit of futurology—wholly inaccurately, as he acknowledges, that the then leftist Labour Party won the 1983 election and removed all American missiles from British soil. The world was engulfed by nuclear war, England emerged unscathed, but the country faced environmental devastation and famine due to the nuclear winter. After a period of lawlessness, the remaining corporations and fascist groups took over Britain and formed a new totalitarian government.

"YOUR POWERS OF OBSERVATION CONTINUE TO SERVE YOU WELL"

Moore never gets on with the film adaptions of his work. Following his negative experiences with *From Hell* (2001) and *The League of Extraordinary Gentlemen* (2003), he rejected the money and credit from Hollywood adaptations. Based on the script, Moore particularly disliked what he saw as a dilution of the anarchy versus fascism theme of his comic, stating it's been: "turned into a [George W.] Bush-era parable by people too timid to set a political satire in their own country . . . [A] largely impotent American liberal fantasy of someone with American liberal values standing up against a state run by neoconservatives."

We'll come back to liberalism later. Certainly though, the film references the War on Terror, which makes sense given its timing, but it's hardly timid (David Lloyd however did endorse the film, calling it "terrific"). Whatever the changes, the regime in the film is still obviously fascistic and hardly "defanged" as Moore claimed. And though he's not quite as violent and indiscriminate as in the comic, the film's V is still a bomb-throwing anarchist (we'll also come back to what that means later). Most importantly, despite Moore's historical predictions being somewhat off, it still remains a possible, even likely, future.

In this at least for the moment fictional future, Norsefire's elite meets with Sutler, his face projected Big Brother style on a large screen. Included are Inspector Finch (Stephen Rea) of the Metropolitan police, Dascomb, who controls broadcasting, Etheridge, who manages the auditory surveillance system, Creedy (Tim Pigott-Smith) from the secret police, and Heyer, who oversees the CCTV network (they make up the nose, mouth, ears, fingers, and eyes of the government respectively, with Sutler sitting at the brain).

Sutler decrees that the destruction of the Old Bailey will be presented to the public as an 'impromptu demolition project' to make way for a new building (the 'new society' shouldn't hang onto the past). And thanks to the regime's surveillance, Evey's identity is quickly discovered. Sutler demands her capture and interrogation.

It's not a cover story the regime wants to have to tell. Once, capitalism seemed intrinsically futuristic (we discussed Hayek's 'market computer' in the previous chapter). The ideas that underpin market economies—growth, accumulation, investment—also express an unspoken assumption: that tomorrow will be different and probably better than today.

But conservatism has changed before our eyes. At first, as we discussed in the previous chapter, it tried to sell a modernized future, better than the one offered by liberals or the left. It's not selling utopia anymore, not even a positive vision of the future, only survival against threats and 'the other.' The reason is the world it's created. As Max Haiven, author of *Revenge Capitalism* (2020), says: "Authoritarianism today does not force society into a formation to fulfil perverse dreams of utopian potential. Rather, it offers austerity, purification and revanchism as means of survival in an ever more hostile world."[2]

"SO DO YOU KNOW WHAT'S GONNA HAPPEN?"

There are really two main types of fictional dystopias: ones where the state collapses and we're left in a postapocalyptic wasteland, and others where a totalitarian regime exploits collapse to seize power. So our 'choice' is postapocalyptic, or postdemocratic.

Perhaps for dramatic reasons, there are many more post-apocalyptic movies and other media than there are realistic collapse stories—among them *The Omega Man* (1971), *A Boy and His Dog* (1975), *Mad Max* and its sequels (starting in 1979), *12 Monkeys* (1995), *Waterworld* (1995), *The Postman* (1997), *28 Days Later* (2002), *The Day After Tomorrow* (2004), *The Road* (2009, based on Cormac McCarthy's 2006 book), *I Am Legend* (2007), *The Book of Eli* (2010), *Snowpiercer* (2013), and *World War Z* (2013). Even some of the most influential dystopian novels have apocalypses in their backstories, from *Nineteen Eighty-Four* to *Neuromancer*.

Prompted by terrorism, the global financial crisis, ecological devastation and climate change, and now the pandemic, as well these fictional depictions, there's been a growing public interest in collapse. For example, in nonfiction books such as Sir Martin Rees' *Our Final Hour* (2003), James Howard Kunstler's *The Long Emergency* (2005), Jared Diamond's *Collapse* (2005), James Lovelock's *The Revenge of Gaia* (2006), Naomi Oreskes and Erik Conway's *The Collapse of Western Civilization* (2014), Roy Scranton's *Learning to Die in the Anthropocene* (2015), David Wallace-Wells' *The Uninhabitable Earth* (2019), Pablo Servigne and Ralph Stevens' *How Everything Can Collapse* (2020), and Fabian Scheidler's *The End of the Megamachine* (2020).

And this is just the most recent wave. As Diamond noted, his book is among a number of contemporary studies of civilizational collapse, for example Joseph Tainter's *The Collapse of Complex Societies* (1988), Ronald Wright's *A Short History of Progress* (2004), Norman Yoffee and George

Cowgill's *The Collapse of Ancient States and Civilizations* (1988), and Charles Redman's *Human Impact on Ancient Environments* (1999). And we've discussed predictions of collapse such as *The Limits to Growth* going back to the 1970s.

This is not to mention the large literary (and often not so literary) genre of postapocalypse novels. Some of the latter are survivalist stories, including those coming from the (often Christian) right, by writers such as the survivalist James Wesley, Rawles (*sic*) and the post-Rapture *Left Behind* series by Tim LaHaye and Jerry B. Jenkins. At the extreme, there are the white supremacist 'novels' that go back to William Luther Pierce's *The Turner Diaries* (1978), often described as the Bible of the racist right.

But some people aren't just consuming the end of the world, they're preparing for it, and this accelerated during the pandemic.

"DEFIANT TO THE END, HUH?"

According to Bradley Garrett in *Bunker: Building for the End Times* (2020), 3.7 million Americans identify as 'preppers,' but research from Cornell University suggests the number of 'extremely prepared individuals' in the United States may be as high as 11.4 million people (defined as someone who could survive for a month or more at home without power, water, or transportation).[3]

Some of this represents a sensible recognition that there's something increasingly fragile about our societies, and the state isn't going to save you in crisis, which is probably true. It can also be interpreted as a rejection of consumerist dependency and for self-sufficiency. As sociologist Richard G. Mitchell puts it in *Dancing at Armageddon* (2002), survivalists "want a part in creating the marketplace, in fashioning economic culture, in crafting rituals of appraisal and exchange" (albeit often from a different politics, this echoes the 1970s environmental consciousness discussed in chapter 4).[4]

But there's a darker side to some survivalism: a longing for apocalypse because they hope it will bring 'purification.' This reflects right-wing libertarianism and sometimes outright racism: no more nanny state, bureaucratic interference, or political correctness, just social Darwinism, a reassertion of 'traditional' gender roles, and white supremacy. They look forward to a post-crisis world in which the 'undesirables' perish or are rounded up—a distillation of the social contract into something more essential and exclusive.

The problem is, despite all the time they spend reading and thinking about it, survivalists often misunderstand collapse. It's possible we could face some apocalyptic event, a sudden and total disaster, in which case this book, and everything, is probably moot. But total collapse is fantastical and can induce passivity rather than action. The strange paradox is that survival is often depicted as individualist, violent, and often ultimately hopeless, so what difference does it make in total collapse to be able to survive for a month or so? Most survivalists' ideas of their own survival are just a fantasy—collapse-themed cosplay.

As climate futurologist Alex Steffen argues, apocalypse is a misleading model for understanding the future.[5] The crises we'll face in the twenty-first century probably won't be the singular, catastrophic world-ending events of the movies or survivalist novels, and the survivalist Elect won't be the only ones left. It certainly won't be an adventure.

Given the business-as-usual stance of mainstream media, at least survivalism and fiction promote thinking about collapse. But depictions of disaster distract from the long emergency—of ongoing failing systems, environmental, economic, social, and political crises—one we're going to have to face, and are increasingly already in. And unlike apocalypse, collapse won't be the end of the story.

Despite its dramatic flourishes, it's the second type of story, like *V for Vendetta*, that are a better guide, especially politically: there'll still be a state, but an oppressive one. And a better understanding of the likely dynamics of collapse helps us to understand why.

"I KILLED YOU 10 MINUTES AGO"

As Dmitry Orlov, the Russian-American engineer and writer on collapse, argues in books such as *Reinventing Collapse* (2008) and *The Five Stages of Collapse* (2013), the United States and other countries are likely to face a Soviet style degeneration—a collapse in faith in business as usual, market-based provision, political and social institutions, and ultimately social trust and kindness. This is due to the overuse of natural resources, soaring debt, corruption, and military overspending. Orlov also emphasizes the personal experience of collapse, and how those who assume they'll be comfortable and secure will also be humbled.

We're already experiencing what in hindsight we'll recognize as an Eastern Europe style collapse, but the capitalist equivalent—a slide into 'slow dystopia,' as the graphic novelist, artist, and filmmaker Sarnath Banerjee calls it. And we're most likely to end up with regimes like that in *V for Vendetta*. After all, 'apocalypse' means revelation, an uncovering of the truth. More likely than irradiated wastelands, collapse reveals the truth of rightwing politics: as a result of neoliberalism, hollowed-out states that are unable to provide for or protect people in crises, replacing this weakness with cult leader authoritarianism.

Firstly, though, we need to recognize that *The Limits to Growth* was right: we're heading towards collapse, and it's unavoidable. In 2008, the predictions of the original model were revisited. The frightening result was that developments over the past thirty years fitted the model: diminishing resources, increasing pollution, slowing growth, and then collapse. As the authors of the study noted, we're already living through a world in overshoot, where we're drawing on the world's resources faster than they can be restored, and releasing waste and pollutants faster than the Earth can absorb or render them harmless.[6]

Over decades, 'we' have squandered the opportunity to change course. It's true that we keep finding more oil and more minerals. The real problem at the moment is not too little oil, but too much, and the pollution it causes. Perhaps not coincidentally, according to another analysis, we exceeded the 'regenerative capacity' of the biosphere—the land and water required to produce the resources consumed and to assimilate the wastes generated by humanity—*in the 1980s*.[7] That was our last chance to largely avoid collapse, but the new right won power and stole the future.

Remember though, we've only just entered the period in which *The Limits to Growth*'s drama is predicted to unfold. None of the report's scenarios suggested we'd encounter significant difficulty before 2010–2030. In many ways, collapse is here ahead of schedule.

There will be no more sequels based on *The Limits to Growth*'s computer model, though. It can no longer serve its purpose, which was as a warning for us to avoid collapse. Starting from today, no plausible assumptions produce any result but overshoot. As Dennis Meadows, one of the study's authors, says, "There is no sense in only describing a series of collapse scenarios." The new right killed us decades ago.

"I STILL DON'T UNDERSTAND IT, WHY THEY HATE US SO MUCH"

The next morning, Evey, who works as an assistant at BTN, delivers an unmarked package to the costume department. Her supervisor opens it and discovers black cloaks and Guy Fawkes masks. V walks into the lobby of the station with a bomb strapped to his chest. Inspector Finch and his partner hurry to the station to apprehend Evey, while V forces most of the TV station to evacuate. Evey hides in a nearby office while V orders the staff to play a message video on the emergency broadcast system. He addresses the public: "[T]he truth is, there is something terribly wrong with this country, isn't there?" V identifies the culprit: the public's fear and how it turned them to the high chancellor, Adam Sutler, with the price of their silent, obedient consent.

V invites the public to join him in one year in front of Parliament, where he promises to give them a fifth of November that will never be forgot.

But this is a regime that's keenly aware of challenges to its rule. The virus, apparently released by terrorists, killed 100,000 people; this was justification for exerting vast powers over the lives of citizens. To keep the country 'safe', Norsefire sought out those they considered an 'ungodly' threat. Political opponents, immigrants, Jews, Muslims, atheists, homosexuals, and other 'undesirables' were imprisoned and executed. 'Resettlement' camps were established across the English countryside, the most famous of which was Larkhill.

As we discussed in chapter 1, cyberpunk persists because it increasingly describes our present. As its name suggests, it tends to focus on, and heroize, the marginalized, the hustlers, the subversives—the people the megacorporations (having replaced the state) largely ignore. But anyone secretly hoping for a 'cool cyberpunk future' is going to be sorely disappointed. As in *V for Vendetta*, fascist states that seize power don't ignore those who are 'different,' anyone who isn't white, middle class, supposedly Christian, and heterosexual. Instead, they persecute them. And anyone who manages to survive, let alone challenge the regime, is going to be severely traumatized, as we'll see.

"NO ONE ESCAPES THEIR PAST"

Moreover, fascism not only uses fear, it's also deeply fear based itself, and the reason conservatism can mutate into fascism is that fear has long been central to its ideology, lurking like a dormant strain that can always come back to life, given the right conditions.

As with the crises we're going to face, the answers lie in the past. Historically, conservatism has long been catastrophe minded: its concern with 'civilizational collapse' because of the perceived decline in values, the revolution in sexual freedoms, the supposed decline of the family, the rise of feminism (and the supposed 'feminization' of men), immigration bringing different 'cultures', and ultimately the 'decline of the West.'

But there's also an internal fear, a genuine one for some conservatives, seemingly—about enemies, threats, conspiracies, and their secret plans and agendas. The right has always had a tendency to view the left in paranoid terms, reflecting its fear of social disorder, and ultimately its fear of the masses—from communism, the 'red menace,' 'anarchy,' new social movements, 'cultural Marxism,' community organizing (yes, really), and conspiracies relating to changing demographics (we'll come back to conspiracism in a moment). This is because conservatives fear becoming a minority—marginalized, persecuted, even eradicated. As a result, at times (one of them being now), the right sees any victories for the liberal/left as potentially apocalyptic. As this fear has become increasingly mainstream on the right, so has the view that no quarter can be given, that every battle is existential, and relatedly, that voting and political participation need to be curtailed.

As we noted in chapter 5, environmentalism and reports such as *The Limits to Growth* added to this anxiety on the right. Conservatives claimed (and still do) that environmentalists intend to end 'your' freedom and the 'American way of life.' Conspiracism centered on environmentalism has been growing ever since, around sustainability, the United Nations' Agenda 21, climate change, the Green New Deal.

Conservatism has always had a self-proclaimed attachment to 'law and order,' but combined with paranoia this starts to take on a different, much darker hue—a rationale for preemptively stopping 'them' before they're able to introduce their dangerous, freedom-destroying programs. It's propaganda and projection, but when you think your enemies are this dangerous, it's only a few steps to justifying their elimination.

Guy Fawkes *was* a conspirator in a plot, of course, and we'll see whether V is a threat to the regime. But in real life, liberals and the left aren't a threat. Honestly, we have no grand project, no secret program, no takeover plan. We don't even have one for our own survival.

And yet, the right's fearfulness about its opponents is not just the raving of madmen—it has intellectual, seemingly respectable roots. This new, more virulent strain of fear was developed in the *university* lab.

Before Hayek, or rather before his ideas were adopted by politicians such as Reagan and Thatcher, post–Second World War mainstream conservatism had been about finding an accommodation with liberalism, even leftism, by slowing progress but with a recognition that it couldn't be stopped entirely. It was this compromise that the new right's revolutionaries rejected as a flawed, and fatalistic, strategy.

As we've seen, Hayek regarded even incremental social programs as a road to serfdom. After Hayek, it was a moral imperative to counter progress, even eradicate it ideologically. For all of his proposed defense of a liberal order, and despite being an academic economist (albeit one who wanted his ideas to change the world), Hayek's paranoia about even gradualist progressive reforms injected a renewed paranoia into conservatism, one that's still mutating today.

These neoliberals often claimed to reject conservativism; Hayek himself wrote an essay in 1960 titled 'Why I Am Not a Conservative', arguing that traditional values are hostile to technology, innovation, and the free market. But really, Hayek was always fundamentally conservative, in his hyperbolic warnings against center-left programs, his fear of equality, 'social designs' and so on, and because as we've seen, the policies he promoted have benefitted the already powerful. Fearfulness is also why depicting neoliberals and the new right as being 'anti-state,' and for freedom and individual liberty as they claim, ignores their extension of the state in many respects—in so-called law and order, the defence of social hierarchy, and especially, now, in their conservative authoritarian endgame.

This is the fascism at the heart of freedom loving neoliberalism. Hayek's road to serfdom will come true, not because we didn't listen to him, but because we did.

"THERE ARE NO COINCIDENCES, ONLY THE ILLUSION OF COINCIDENCES"

As V's broadcast ends, the police try to capture him, but V has forced everyone in the TV studio to wear a Guy Fawkes mask. V escapes, taking an unconscious Evey with him.

Evey wakes up in V's lair. She asks to be released, but V tells her she must stay until his goal to destroy Parliament is completed, otherwise she'll be captured by the regime and forced to give them information about him. He explains that his plan to destroy Parliament is not just to bring down a building, but to instill the idea in people—that their government should be afraid of them, not the other way around.

Later, V kills TV pontificator Prothero in his home. Investigating the murder, Finch discovers that Prothero used to be an officer at Larkhill. Evey sees the announcement of the murder on the news. Evey questions V about it, who states he plans to kill others in the name of 'justice.' Secretly, Evey resolves to escape; she will offer her assistance and look for an opportunity to get away. But V agrees to let her help and comes up with a plan for his next victim, a pervert bishop who's also a senior party member.

Evey goes along with the plan, but escapes. She finds sanctuary with her boss, Gordon Deitritch (Stephen Fry). They watch the latest episode of his TV show, which includes a satire of the regime. Evey is worried that Gordon will face repercussions, but he tells her not to fret. As a wealthy, white man (albeit also closeted gay), he's protected. However, that night, soldiers led by Creedy break in, place a black bag over Gordon's head and take him away. Evey manages to climb out the window, but she's also apprehended and bagged before she can escape. Like Gordon, the liberal-minded Evey thought she might be able to find a way to live under the regime. Of course, she, and he, and many others, were wrong. She's forced to face not only the nature of the regime, but next, what needs to be done to confront it.

Conservatism's paranoia about its opponents also creates the perfect petri dish for other dangerous conspiracy theories, something that's increasing in early collapse, as crises bring new opportunities for blaming, division, and containment.

Rather than being a freak, inexplicable occurrence, conspiracy theories such as QAnon are an example of this longstanding paranoia, and exemplify the hope that liberals will be rounded up, arrested, detained, and executed. It's probably just the start. Ironically, given their dismissal of 'environmental doom', apocalypsism is overtaking the right.

Of course, many of the authoritarian populists and opportunists pushing conspiracy theories aren't stupid. They don't really believe that climate change is a Chinese hoax. But it works for the moment. They target the disillusioned and alienated, and give them targets to blame for how they feel. Such theories also provide a pipeline to white nationalism.

As Naomi Klein has demonstrated, in the past neoliberals have used crises to advance their economic agenda (deregulation, privatization, and so on). Today's, and tomorrow's, crises, seized on by conservative conspiracism, are likely to be used for even more destruction, since as Gregory Claeys Claeys (*Dystopia: A Natural History*) suggests, one of the characteristics of political dystopias is that they are closed groups with a paranoid view of those outside the group as a threat to be controlled, removed, and finally eradicated. And as we know from history, paranoid ideology can easily turn into actual extermination.

"THERE'S SO MUCH CHAOS, SOMEONE WILL DO SOMETHING STUPID"

One of the iconic moments in *V for Vendetta* is the domino scene, where V tips over a huge display of black and red dominoes to form a giant letter 'V' (filming it involved 22,000 dominoes; apparently it took four professional domino assemblers 200 hours to set them up.) It comes at a point where V is clearly unsettling the regime and his agenda seems to be gathering public support.

We might like to believe that increasingly obvious economic and environmental crises will cause a change in public consciousness that political leaders will have to respond to. But it's much more likely that such crises will lead to division and repression. In many ways, the crises here presently have already begun turning our politics in this direction.

Globally of course, it *is* a postapocalyptic *Mad Max* world for many people. As we deplete natural resources (we're using twice as many resources as in 1980), this is causing more conflicts over fossil fuels, water, food, and land—more than 2,500 conflicts worldwide are directly related to the ongoing collapse of biodiversity.[8] Climate change alone already causes an estimated 300,000 deaths every year and affects at least 300 million people.[9]

This is driving mass displacement and migration, and triggering nationalistic responses. For example, by creating severe droughts which drove and exacerbated conflicts, climate change was a significant cause of migration and asylum seeking in the first half of the 2010s. As many as one million desperate refugees turned up on European shores; this played a pivotal role in the UK campaign to leave the European Union and the revival of nationalist sentiment across Europe, the US, and elsewhere.

Among many other studies, the same computer model that demonstrated this causal link (developed by the Global Sustainability Institute at Anglia Ruskin University) predicts the collapse of the global food supply system and the virtual collapse of industrial civilization by 2040 due to climate change, water scarcity, energy crisis, and political instability.[10] In this sense, as Robert D. Kaplan has argued, despite being derided by so many commentators for so long, Thomas Malthus was essentially right: population increase, urbanization, *and* resource depletion are undermining fragile governments across the developing world, and also represent a threat to the developed world.[11]

As Thomas Homer-Dixon, founder-director of the Cascade Institute, a Canadian research center, who uses complexity science to examine threats to global security, suggests, the more dissatisfied and afraid people become, the more they tend to cling to in-group (religious, racial, or national) identities. As Homer-Dixon says, "You're setting up the psychological and social prerequisites for mass violence."

As poorer nations disintegrate, waves of migrants will stream out of failing regions, seeking refuge in more stable states. Western societies will respond with restrictions and bans on immigration, multi-billion dollar walls, border-patrolling troops and drones, heightened security, and more authoritarian populist governments. Christian Parenti calls this the politics of the armed lifeboat, "responding to climate change by arming, excluding, forgetting, repressing, policing and killing." It's even possible to envisage an eco-fascist future, taking some elements of environmentalism, for example concerns about population and resource use, to 'justify' exclusion and violence against 'the other.'

It's for these reasons, as well as the scale and cost of the great transition required to a sustainable society (as we'll discuss in the next chapter), that centrist, center-left, and left notions of how a transition might be achieved are often highly unrealistic. They assume a stable state, with sufficient resources, able to lead such a transition. But in a context of fear and scarcity, where the right is busy blaming and othering and conspiracizing, establishing and maintaining a broad political coalition involving sacrifice over many years will be extremely difficult, if not impossible. Partly as a result of neoliberal propaganda, the center-left has struggled to persuade voters that it can deliver even much more modest programs. And yet we're asked to believe that a much larger and more profound change could be within our grasp as crises escalate.

The dominos are starting to fall. The problem is, they're not falling towards justice and equality, but into a fear-fueled, fascist future.

"YOU ALREADY HAVE THE INFORMATION. ALL THE NAMES AND DATES ARE INSIDE YOUR HEAD. WHAT YOU WANT, WHAT YOU REALLY NEED, IS A STORY"

Despite being warned off, Finch continues his investigation, focusing on Larkhill. He discovers the bishop also used to be stationed there. Only one high ranking official connected with the camp is still alive, a woman named Delia Surridge (Sinéad Cusack). But once again, V has gotten there first. At least she apologizes for her actions before V kills her painlessly.

Finch finds Surridge's notebook, which records her work at Larkhill. She was testing drugs on prisoners. All the tests proved fatal, except in the case of one man who was kept in room five. He developed almost superhuman strength and reflexes. On November fifth, the laboratory exploded. As Surridge escaped, she saw the man, horribly burnt and roaring in anger. Finch presents the notebook to Sutler, who dismisses it.

Finch is then contacted by a man named William Rookwood, who tells him the whole tale—starting with a young up-and-coming politician, a deeply religious man and a member of the Conservative party, who seized the opportunity of the St. Mary's crisis to take power (some believed that it was the work of God), and subsequently to appoint himself to the newly created position of high chancellor.

In reality, Sutler, who had been the Secretary of Defence, produced a deadly virus from the only Larkhill prisoner who survived. It was V, of course. Sutler secretly released the virus in a school (after which the virus is named), a subway station, and a water treatment plant. The virus killed more than 100,000 people (in the movie, this figure is meant to be shocking). Norsefire used the public panic to win an overwhelming majority in Parliament and elevate Sutler to High Chancellor—as well as profit off the cure through the leadership's ownership of pharmaceutical companies.

Later however, Finch discovers that Rookwood has been dead for some time; it was V in disguise. Finch struggles to believe the story, but his faith in the regime is shaken. As Finch says: "I suddenly had this feeling that everything was connected. It's like I could see the whole thing, one long chain of events that stretched all the way back before Larkhill. I felt like I could see everything that happened, and everything that is going to happen . . . And I realized we're all part of it, and all trapped by it."

In our world, it doesn't really make sense to call something that's been hidden in plain sight all along a 'conspiracy.' Certainly, there have been conspiratorial aspects to the collapse we're facing: the reaction to studies such as *The Limits to Growth*, the oil companies that buried evidence that carbon emissions would radically destabilize the atmosphere, the organized, well-funded denial, the 'research' from corporate sponsored university economics departments that 'proved' that policies that helped the rich and powerful would benefit everyone, and so on.

The right didn't explicitly decide 'we're deciding to doom the planet.' But they emphasized boundless growth, promised prosperity, dismissed 'dangerous radicals' who want to make you poorer and control your life, campaigned against 'unnecessary regulations,' promoted the idea that magical future technologies would solve all our problems, celebrated rampant consumerism, and sought to marginalize the 'threat of democracy' from trade unions, campaigners, new social movements, and others.

Everything since has brought collapse ever closer, and the worst kinds of collapse leading to dystopia, the kinds explored in this book—whether a cyberpunk, corporate-controlled future, Biff Tannen–style private affluence and public squalor, environmental postapocalyptic style disaster, or in this case, a corrupt authoritarian theocracy.

Not only have conservatives denied the approaching crises until it's too late, perhaps not coincidentally they're much more prepared for how the endgame plays out politically. They attacked government and promoted competitive individualism, severely hampering our ability to address crises collectively. They proclaimed the end of history, closing off alternative futures. They hugely increased inequality, generating widespread anger and resentment. They spread post-truth politics, disinformation, and confusion, by undermining science, expertise, and objectivity. They propagandized their supporters against 'elites' and their 'hidden agendas.' They used crises as an excuse to make the non-rich even less secure. They blamed minorities and migrants. They eroded and corrupted democracy. They strengthened corporate control, and attacked international institutions and agreements. And then they offered up authoritarian leaders promising to 'master the chaos' they'd created.

We were told what the result would be by the Cassandras who tried to warn us, and through dystopian stories. Like Finch, we know it's true, even if we struggle to accept it. We already had the information. What we needed was the story.

"IF OUR OWN GOVERNMENT WAS RESPONSIBLE FOR THE DEATHS OF ALMOST A HUNDRED THOUSAND PEOPLE . . . WOULD YOU REALLY WANT TO KNOW?"

In 1980, the English historian and antiwar activist Edward Thompson proposed the word 'exterminism' for "those characteristics of a society . . . which thrust it in a direction whose outcome must be the extermination of multitudes."

At the time, anti-nuclear campaigners regarded political leaders as having succumbed to madness. Thompson tried to answer the question of how this fits with capitalism's self-interest.[12] As Thompson suggested, conventional Marxist analysis doesn't explain it. Of course, radical thinkers have long noted the links between capitalism and the politics of killing. Fredrich Engels used the powerful phrase "social murder" to describe how political elites coldly embrace policies that ensure "early and unnatural" deaths for the impoverished and marginal. But nuclear madness amounted to collective suicide.

The best explanation Thompson could come up with was that exterminism—the obvious irrationalism of nuclear weapons and mutually assured destruction—was the inevitable outcome of economies which relied on profits from arms manufacturing, heavily influenced by the military-industrial complex and bureaucracies jostling for power and influence. It was the *interests* underpinning the whole system. So it is with the broader collapse we've been discussing here. As the new left warned us, it's the way the machine works.

Exterminism was used by Peter Frase in his book *Four Futures: Life After Capitalism* as the name for the most likely, and bleakest, of the futures we face: a dystopia in which the rich retreat to fortified compounds where robots do all the work, while the rest of us drown in the hell of a rapidly warming planet (much like the 2013 sci-fi film *Elysium*).[13]

As Frase suggests, the right's endgame is not to save civilization, but to save a capitalist order that protects and privileges the wealthy and powerful. Historically, capitalism has needed workers to keep its factories and shops going, but this was always an 'inconvenience' to the ruling class. As Frase notes, this is where automation can help, allowing for the slow extermination of 'superfluous' multitudes, leaving the rich to quiet luxury inside their utopias.

Exterminism is the ultimate story of the dystopias 'we've' been creating. It's the secret hidden in plain sight, which is what a lot of dystopian films are really about, asking us 'don't you see what's about to happen?' Our weakness, in a way, is that the center/left wants to protect people. The right is increasingly happy to see them suffer and die.

If 'exterminism' sounds paranoid, these are the decisions we already make but prefer to pretend we don't: who has shelter, water, heat, food, medicines, security, and who doesn't. Often, these decisions are rationalized in cool, dry academic analysis, and presented as perfectly reasonable. So it will be in the endgame.

One field in which exterminism has been apparent is economics. It was there, barely hidden, in neoliberal thinking—an economic exterminism, deciding on the basis of abstract mathematical models which industries, areas, and people are to be rendered redundant.

It's also been there in the dismissal of environmental concerns. As we noted in chapter 4, up to the 1960s, most conservatives were broadly supportive of protecting the environment. But in reaction to the rise of new left environmentalism, the libertarian conservatives associated with the University of Chicago's departments of economics and law increasingly opposed environmentalism. Economists like Milton Friedman, George Stigler, and Ronald Coase rejected the idea that polluters should take responsibility for the social costs of their actions. Pollution signified a booming economy, and any environmental problems would be more than offset by the gains from growth. Corporations' only responsibility is to make money for their shareholders. As their influence grew, these views became mainstream on the right.

Such arguments have not only accelerated the crises we face, in collapse they're dangerously exterministic. For example, in the 1990s, the Yale economist William Nordhaus invented the first integrated assessment models to explore how growth affects carbon emissions and vice versa. The United States, and to some extent the Intergovernmental Panel on Climate Change (IPCC), relies on a model developed by Nordhaus, for which he was awarded the Nobel Prize in 2018. Which sounds good, except that it shows that if we rapidly reduce carbon emissions in line with what scientists say is necessary, this would significantly slow down growth. Nordhaus' career has been devoted to finding what he calls a "balance" between climate mitigation and growth. In a famous 1991 paper, 'To Slow or Not to Slow', he argued firmly for the latter.

Like the Chicago school, Nordhaus' argument heavily 'discounts' (in economic language) the lives of future generations by claiming that even if it locks in more climate change, higher growth now will make these future generations richer and so more able to manage its consequences (again, this partly relies on the idea that as-yet-invented technologies will magically resolve our problems for us). Using this logic, Nordhaus claimed it's 'optimal' to warm the planet to about 3.5 degrees Celsius over preindustrial levels—catastrophically higher than the 1.5 degrees Celsius limit that the IPCC calls for.

This is a collapse agenda. It's absurd to believe that the global economy would just keep going despite a massive failure in the world's food supply, for example. Climate chaos will displace, starve, and kill millions of people, yet the implicit assumption is that most of them will be poor, non-white, and non-Western. Such people don't 'contribute' much to growth anyway, so it's a price worth (them) paying, apparently.

So it's also seemingly white supremacist, the valuing of some lives more than others, the assumed, sometimes unconscious entitlement that some people have a right to survive—not just survive but to live lives of luxury—while others can be left to perish. It's only a few steps from this to an explicitly white supremacist regime that marginalizes, marshals, and eliminates those lives deemed to be worth less, or worthless.

There's a hellish circularity to it as well. An implicitly white supremacist approach to crises (for example, climate change) helps to create explicitly white supremacist regimes. Crises induce the fear that opportunistic authoritarians seize on, and serve to legitimize their regimes. It's why authoritarianism, racism, and hierarchy are all connected. In turn, 'our' material and geographical advantages at the expense of others are then used to reinforce white supremacism—recall Norsefire's propaganda about the 'British character' as special and destined to "prevail."

Lone survivalists can entertain fantasies about their own chances of post-collapse living all they want. The only survivalists who really stand a chance are elites, and they've been preparing for some time. As Sarah Kendzior puts it: "In the eyes of autocrats and plutocrats, the future is not a right but a commodity. As climate change brings unparalleled crises, the future becomes a rare asset, meant to be hoarded like diamonds or gold."[14]

"I WANT ANYONE CAUGHT WITH ONE OF THOSE MASKS ARRESTED!"

When the bag is removed, Evey finds herself in an interrogation room. A government official, hidden in darkness, asks her about the location of 'the terrorist.' When she can't answer, she's 'processed.' Her head is shaved, she's tortured, and kept in a small cell.

One day, she hears a noise from inside one of the walls and discovers a note scribbled on toilet paper. It's from a woman named Valerie, who tells the brief story of her life. Valerie describes how she first fell in love with a girl in her class. Later, she met the love of her life, Ruth, and they moved to London as America's civil war worsened and the British government began to 'change.' Ruth was 'taken away' shortly before Valerie. She ends her note by offering some solace that while her life may be taken, she still retains her integrity as a human being, and tells her reader that, though she doesn't know them, she loves them.

Tearful but inspired, Evey refuses to cooperate with her captors. She's returned to her cell to await execution. An official arrives and pleads with Evey to provide just one bit of information to save her life, but Evey says she'd rather die. The man then tells her she now has no fear and is completely free, then leaves. Confused, Evey gets to her feet and walks out the now open door. She goes down a corridor and turns a corner to see that she's in the Shadow Gallery. V explains that her imprisonment was an illusion designed to make her let go of fear and realize there's something more important than her life. Evey collapses. V takes her to the roof. Evey steps out into the rain and raises her arms, laughing in a moment of pure release, just as V once did.

Young people in many countries are becoming radicalized. And sometimes the *V for Vendetta* Guy Fawkes mask has cropped up as a symbol of defiance, from the demonstrations at the G20 and G8 conferences in London and L'Aquila in 2009, the Occupy protests from 2011 starting in New York and spreading around the world, the Arab Spring including the 2011 Egyptian revolution, the Hong Kong protests in 2014 and 2019, demonstrations in Thailand, Myanmar, Turkey, Brazil, and others, and the computer hacking group Anonymous (the mask has been banned in some Middle Eastern states).

What these young people have realized is the true nature of the regimes they face, and they've been prepared to sacrifice their freedom and in some cases their lives. But as the Sutlers already lurking in the darkness spy their chance, will more of us be prepared to be so brave, and before it's too late?

This is dramatized in *V for Vendetta*. Sutler rails at his subordinates: why haven't they apprehended the terrorist? As happens under real regimes, he issues stricter curfews and surveillance on civilians who, he's angered to hear, gossip about V and how they believe he's still alive. And he is. V visits Creedy and makes him an offer. He tells him that once V destroys Parliament, Sutler will have no use for Creedy and will sacrifice him as a scapegoat. If Creedy gives up Sutler, Creedy will be promoted and he'll be able to deal with V as he chooses.

As the fifth of November nears, thousands of crates are delivered across London, each one containing a costume: a black cloak, hat, and Guy Fawkes mask. Sutler decrees that anyone caught wearing them will be arrested. Riots break out across the city. A young girl in a V costume is shot by a Fingerman, but despite displaying his Norsefire badge, he's beaten by a vengeful crowd. People are no longer prepared to cower before the regime.

Evey returns to the Shadow Gallery. V shows her an abandoned subway tunnel underneath the building. He's filled a train filled with explosives, which V says will go along its track to Parliament. He leaves it to Evey to decide whether to use it. Meanwhile, he'll go to meet his fate. Evey begs him to stay, but V disappears into the tunnel.

Sutler gives a speech vowing swift justice against anyone caught outside after curfew. But no one's listening anymore. Creedy keeps his deal with V; he captures Sutler and shoots him in the head. Then he demands V remove his mask. When V refuses, Creedy orders his men to open fire. V is hit by a barrage of bullets, but kills all of the soldiers. Creedy asks why he won't die. V tells him that beneath his mask is more than flesh, "there's an idea, and ideas are bullet-proof." V breaks Creedy's neck, before staggering back down the tunnel. He removes a large metal chest plate, but he's bleeding profusely. Evey has been waiting for him. V confesses he's in love with her, then dies.

Soldiers surround Parliament but can't get in touch with anyone in authority. They see a massive crowd of people advancing towards them, all clad in Guy Fawkes costumes. Overwhelmed, they put down their weapons and allow the crowd to walk past them.

Finch finds his way into the subway and discovers Evey at the controls to the train. She's placed V on a seat inside the train, arms crossed, covered in roses. She tells Finch the country needs more than a building now, it needs hope, as she pulls the lever and steps out of the train. Disillusioned with the regime, Finch doesn't stop her. Evey leads Finch to the roof where they have a clear view of Parliament. In front of them, with the crowd convened, Parliament explodes to a soundtrack of Tchaikovsky's *1812 Overture*.

Finch asks who V was. Evey says, "[H]e was my father and my mother, my brother, my friend. He was you and me. He was all of us."

"THIS VISAGE, NO MERE VENEER OF VANITY, IS A VESTIGE OF THE VOX POPULI, NOW VACANT, VANISHED"

Earlier, V had explained his belief in the power of symbols and his adoption of the Guy Fawkes persona. His mission was to revive a public spirit of defiance. It's just a film, of course, a fantasy. Maybe we would rebel en masse. Some regimes do fall that way. But the protests and revolutions we mentioned earlier largely failed—the authorities clamped down, and arrested, imprisoned, and 'disappeared' the protesters. Perhaps freedom and justice are ideas that will never die. But that doesn't mean we'll be able to bring down any future fascistic regime, certainly not so easily. And in England, Guy Fawkes Day festivities are not celebrations of the plot, but its failure.

If we're being honest, it's entirely possible that many people, including self-proclaimed liberals, would weakly accept fascism as they've done before, even as their friends and neighbors start to disappear. And as we've noted, there's another other purpose to dystopian fiction: not just warning us about a possible future, but highlighting the dystopia in our midst, the one that's already here. As Rob Urie challenges Americans to consider, "[W]ould they recognize a fascist government if they were its beneficiaries?" – in a country that already has the largest prison system in human history, increasingly militarized police forces, an oligarch class that has assumed de facto control of the state.[15]

To believe anything else is the fantasy. As Samantha Riedel puts it: "V for Vendetta is not merely a cautionary tale for our country's political morass or trajectory—it is a retelling of events that have, in their fundamental essence, already occurred."[16]

V for Vendetta is certainly subversive, even at the risk of seeming adolescent. Just a few years after 9/11, it imagines a terrorist as people's hero, blowing up landmarks and government buildings, and depicts the fascistic regime that tries to stop him using echoes of the War on Terror (which at the time, still retained much public support).

But its filmmakers also seem to recognize that V reflects the dictators he seeks to depose. He's monomaniacal, spreads fear, foments disorder, and uses torture in the name of his 'righteous cause,' one born, as the title suggests, from a desire for revenge.

Of course, fascists love violence, and in some ways V is a manifestation of fascist violence, as 'superheroes' often are (Alan Moore explored this theme further in *Watchmen*). Fascists want their opponents to use violence so they can justify their repression. They love the idea of turning us into the monsters they are, validating their fears about enemies who need to be eliminated in the name of 'security.'

Obviously, fascism needs to be resisted and defeated. But by using anger and vengefulness, or by finding some other way? As the late David Graeber, the anarchist scholar who coined the 'We are the 99 percent' slogan of the Occupy movement, commenting on *V for Vendetta* said: "Anarchy is about creating communities and democratic decision making. That's what is absent from Hollywood's interpretation."

As we'll discuss in chapter 9, instead of violent fantasies, we should think seriously about how we might survive, and even build our own structures for freedom, justice, and security. In the end, it's Valerie's message that's a better guide. She writes that she knows she will die in prison: "Every inch of me will perish. Every inch but one." The part of her that might survive, fragile though it is, is the only thing that matters—her love, her humanity, her hope.

NOTES

1. Joshua Tyler, "V For Vendetta," *CinemaBlend*, May 27, 2016. www.cinemablend.com/reviews/V-Vendetta-1472.html.

2. Max Haiven, "Capitalism's Vengeful Utopia: Unpayable Debts from Above and Below," *L'Internationale Online*, May 10, 2020. www.internationaleonline.org/research/real_democracy/131_capitals_vengeful_utopia_unpayable_debts_from_above_and_below/.

3. Chris Ellis, *The Noah Virus: Who is Infected with High Resiliency for Disaster?*, August, 2020. www.researchgate.net/publication/343363026_The_Noah_Virus_Who_is_Infected_with_High_Resiliency_for_Disaster.

4. Richard J. Mitchell Jr., *Dancing at Armageddon: Survivalism and Chaos in Modern Times* (Chicago: University of Chicago Press, 2002).

5. Alex Steffen, "The Apocalypse Makes Us Dumb," *Alex Steffen*, May 11, 2015. www.alexsteffen.com/apocalypse_makes_us_dumb.

6. Graham Turner, *A Comparison of the Limits to Growth with Thirty Years of Reality*, CSIRO Working Paper Series (Canberra: CSIRO Sustainable Ecosystems, 2008); Donella Meadows, Jorgen Randers, and Dennis Meadows, *Limits to Growth: The 30-Year Update* (White River Junction, VT: Chelsea Green, 2004).

7. Mathis Wackernagel, Niels B. Schulz, Diana Deumling, Alejandro Callejas Linares, Martin Jenkins, Valerie Kapos, Chad Monfreda, Jonathan Loh, Norman Myers, Richard Norgaard, and Jørgen Randers, "Tracking the Ecological Overshoot of the Human Economy," *Proceedings of the National Academy of Sciences* 99, no. 14 (July 2002), 9266–9271. www.pnas.org/content/99/14/9266.

8. Intergovernmental Science-Policy Platform on Biodiversity and Ecosystem Services (IPBES), *Global assessment report on biodiversity and ecosystem services of the Intergovernmental Science-Policy Platform on Biodiversity and Ecosystem Services* (Bonn: IPBES Secretariat, 2019).

9. Christian Parenti, *Tropic of Chaos: Climate Change and the New Geography of Violence* (New York: Nation Books, 2011).

10. Lloyd's, *Food System Shock, The Insurance Impacts of Acute Disruption to Global Food Supply* (London: Lloyd's, 2015).

11. Robert D. Kaplan, "The Neo-Malthusian World of the Coronavirus, How the Natural Environment is a Force for Instability," *The National Interest*, February 28, 2020. https://nationalinterest.org/blog/buzz/neo-malthusian-world-coronavirus-127947.

12. Edward Thompson, "Notes on Exterminism, The Last Stage of Civilization," *New Left Review* 121 (May/June 1980): 3–31.

13. Peter Frase, *Four Futures: Life After Capitalism* (London: Verso, 2016).

14. Sarah Kendzior, *Hiding in Plain Sight: The Invention of Donald Trump and the Erosion of America* (New York: Flatiron Books, 2020), 2.

15. Rob Urie, "Economic Decline and the Threat of Fascism," *Counterpunch*, May 1, 2020. www.counterpunch.org/2020/05/01/economic-decline-and-the-threat-of-fascism/.

16. Samantha Riedel, "How V For Vendetta Predicted America's Descent into Fascism," *them.*, November 9, 2018. www.them.us/story/v-for-vendetta-fascism-america.

Chapter 8

"Bloody Paperwork"

How Brazil *Dashes the Hopes for a Luxury, Leisure-Filled Future Communism*

A further British-set dystopia, *Brazil* (1985), offers another critique of the promise of a perfectly run technological future—in the case of this chapter, a dream coming predominantly from the left rather than the right. Despite its attractiveness, the recent resurgence of interest in left accelerationism or automated communism, that we could use new technologies to finally realize the perfectly fair, planned economy, not only ignores the very real environmental collapse we're heading towards, it also has worrying implications for democracy. This is where, well before its time, *Brazil* depicts a society overrun by centralized bureaucracy where nothing really works as it should, but the system can never, ever admit it.

"LISTEN, KID, WE'RE ALL IN IT TOGETHER"

It's impossible to think about the film without its theme song entering your head. *Brazil* doesn't take place in Brazil. Rather, it's named after its recurring theme *Aquarela do Brasil* (Watercolor of Brazil), written by Ary Barroso and known as just 'Brazil' to Western audiences (the film uses the version performed by Geoff Muldaur with new lyrics in English by songwriter Bob Russell). But while its use seems nonsensical at first, especially in a dystopian story, it makes sense the more you think about it.

Although not a hit initially, *Brasil* has become one of the twenty most recorded songs of all time. It's fun, joyful, expressive, infectious, and absurd. Barroso wrote the song while he was stuck inside his house during a thunderstorm. It's a dream of escape—one day. Apparently, *Brazil*'s director Terry

Gilliam (we discussed his *12 Monkeys* in chapter 3) got the idea to use the song while location scouting in Port Talbot, a steel town in Wales. He saw a man on the beach watching the sunset and listening to the song on his radio. 'Brazil' is an enchanted, faraway place—of freedom, self-expression, creativity, love, heroism, and adventure (curiously, in Irish myth, Brasil/Hy-Brasil or other variants, is also a phantom island said to lie in the Atlantic Ocean west of Ireland, one which can never be reached). It represents all of the things *not* in this fictional dystopian future-present, one our hero, Sam Lowery (Jonathan Pryce) secretly dreams of escaping, and we will too.

We start on a retro-styled television, showing an advert for Central Services, with its annoyingly jaunty jingle, "We do the work, you do the pleasure." In a fake sincere tone, a middle aged, blazered, Bryl-creamed salesman wants to talk to us about ducts (yes, the tubes used in heating and ventilation). "Do your ducts seem old-fashioned?", he asks us.

As he drones on, the camera pulls back to reveal we're in front of a shop window. All the TVs are playing the same advert. Apparently, "new duct designs are now available in hundreds of different colors to suit your individual tastes." We are encouraged to hurry to our nearest Central Services showroom, "while stocks last." Suddenly a huge bomb explodes, shattering the window and blowing the TV sets out into the street.

As the wreckage burns, one of the TVs plays a fawning interview with the Deputy Minister in the Ministry of Information, Eugene Helpmann (Peter Vaughan). The interviewer asks politely, "What do you believe is behind this recent increase in terrorist bombings?" Helpmann (surely named ironically) replies grumpily: "Bad sportsmanship. A ruthless minority of people seems to have forgotten certain good old-fashioned virtues. They just can't stand seeing the other fellow win. If these people would just play the game, they'd get a lot more out of life."

We're in a run-down, polluted, consumerist, hyper-bureaucratic totalitarian future Britain (again), "Somewhere in the 20th century" we're told by the title card. Welcome to *Brazil*.

"HAS ANYBODY SEEN SAM LOWRY?"

Cut to a busy, seemingly subterranean office. Reams of paper are carried hurriedly back and forth by pullover clad young clerks, while uniformly grey suited older workers rush around with files and type instructions into hulking (also grey) machines. Mr. Kurtzmann (Ian Holm) looks out over this bureaucratic hive of activity, then checks his watch. The moment he shuts his door behind him, his staff huddle around their tiny TVs to watch old cowboy movies, then quickly return to work when he sticks his head out suspiciously to make sure they're still getting on with their jobs.

The always nervous Kurtzmann is also distracted, though. His most trusted confidante is not in the office. Instead, Sam, a mild mannered, low-level functionary in the Ministry of Information (Records Department), is still in his dreams, specifically his recurring dream of soaring through the clouds as a winged warrior to save a beautiful mystery woman (Kim Greist). Much to Kurtzmann's relief, since he relies on him, Sam has no interest in being promoted, seemingly happy in his quiet life in this most dead-end of bowels of the Ministry.

But as we'll see, the system doesn't work, and by inadvertently challenging its dysfunction and by dreaming of something better, Sam (Pauline Kael described him neatly as a "Winston Smith-Waiter Mitty") will soon find himself becoming an enemy of the state.

Kurtzmann is right to be worried. Shortly before Christmas, a fly gets jammed in a teleprinter, so misprinting a copy of an arrest warrant, which results in the arrest and 'accidental death' during interrogation of a cobbler, Archibald Buttle, instead of a suspected terrorist, Harry Tuttle (we'll get to him). Buttle's heart condition didn't, of course, appear on Tuttle's medical files provided to Information Retrieval. It's a mistake!

Gilliam says *Brazil* isn't science fiction, more of a "documentary" or "political cartoon" of the world we already live in. The American-born Gilliam witnessed police riots in Los Angeles in the mid-to-late 1960s, and personal interactions with cops left him no lover or respecter of authority. Soon after, he moved permanently to England (if he hadn't left the United States, Gilliam says he might have become a "full-time, bomb-throwing terrorist" himself).

Gilliam became a British subject in 1968 and renounced his American citizenship in 2006. He became a writer and animator on the British TV children's program *Do Not Adjust Your Set*, where he met many of the members of what would become Monty Python's Flying Circus. Gilliam created the surreal, funny animations for Python, then co-directed *Monty Python and the Holy Grail* (1975). Two years later he directed his first solo feature, *Jabberwocky* (1977), followed by *Time Bandits* (1981) and many other inventive, occasionally future-set stories, often exposing absurdity, particularly that of authority.

As we noted in the previous chapter, there are two main types of (future) dystopian stories, the postapocalyptic or the postdemocratic. *Brazil*'s Britain is certainly the latter. But often, such settings, and the histories of real-life authoritarian regimes, overstate their (ruthless) efficiency. In reality, as in *Brazil*, they tend to be hugely bureaucratic and inefficient, as anything run along ass-covering lines typically is.

Inevitably, machines malfunction, and people make mistakes. The problem is that this is a society that's based on trusting the machines, trusting the system, and covering up awful failures when something goes wrong, because otherwise those at the top of the system fear the whole thing would collapse (they might be right). The system admits no error.

Which begs the question, when presented with yet another dream of a perfectly run technological future—in the case of this chapter, coming predominantly from the left rather than the right—why wouldn't it turn also into a nightmare, as it always has before?

"WHAT IS THIS MESS?"

Brazil was written by Gilliam, Charles McKeown, and Tom Stoppard. Gilliam developed the story and wrote the first draft with Charles Alverson, the co-writer of *Jabberwocky*. Alverson was paid for his work but was ultimately uncredited (Gilliam long denied that Alverson made any material contribution, but later had to acknowledge he had).

Critics were initially divided. As is often the case with Gilliam, it's complicated, as we'll explain later. To Richard Corliss in *Time* magazine, "There is not a more daft, more original or haunting vision to be seen on American movie screens this year . . . A terrific movie has escaped the asylum without a lobotomy. The good guys, the few directors itching to make films away from the assembly line, won one for a change."

In the *Los Angeles Times*, Kenneth Turan described *Brazil* as "the most potent piece of satiric political cinema since Dr. Strangelove." Our friend Pauline Kael was more mixed. As with *Blade Runner*, she recognized it, visually at least, as an original, bravura piece of moviemaking. Yet she didn't find it funny, and thought that its attempts at humor undermined its point: "What Gilliam appears to be saying is that reality is so intrinsically awful that fantasy will always be needed. But fantasy itself, in Gilliam's hands, seems something you want to escape from. This image-maker leaves you with nothing but images."

To which Gilliam characteristically responded, "You're full of shit, Pauline." And he's right, *Brazil* is much more than merely images (it's also very funny). Its totalitarian setting and descent into darkness suggests how a system that's supposed to provide for our wants and needs spirals out of control, how institutions become our oppressors, how we're in collective denial about collapse, but how some people continue to dream of a more human, more humane society.

It's why Salman Rushdie describes *Brazil* as being set in a "cancelled future."[1] Like *Blade Runner*, *V For Vendetta*, and Gilliam's *12 Monkeys*,

Brazil is a retrofitted future-present. This isn't the slick, invisible technology future we were promised. It's a malfunctioning mess. In its confounding, intrusive mazes of ducts (of course), just about held together by make-shift patches, mirroring the system as a whole. In Sam's cramped, hardly leisure-enhancing apartment, with its dysfunctional Rube Goldberg (or Heath Robinson, if you're British) devices that pour coffee all over his breakfast toast. In the tiny TV screens that require bulky magnifying lens. In robotic 'helpers' that get in the way. And in temperamental teletype machines that lead to the death of innocent citizens.

As in *Blade Runner*, there's pollution, towering brutalist architecture, claustrophobic, chaotic streets, advertising as a continual backdrop, and simulations standing in for real things (in a restaurant scene, colored mush is placed next to pictures of the real food it's meant to taste like, and the elite ladies who lunch are obsessed with plastic surgery). And again, ordinary people are powerless in the face of institutions, but this time, it's the state.

In another nod to cyberpunk—*Blade Runner* had been released only a couple of years before—there's also *Brazil*'s film noir styling and lighting. And like *Blade Runner*, *Brazil* was unsuccessful in its initial North American release; it was more successful in Europe and has since become a cult, and highly regarded, iconic film. It also experienced a major struggle over its ending, which we'll get to.

As Jack Mathews, film critic and the author of *The Battle of Brazil* (1987) documentary and book, describes it, the film satirizes the "bureaucratic, largely dysfunctional industrial world that had been driving Gilliam crazy all his life." As Gilliam reflects about *Brazil*, together with his 2013 feature *The Zero Theorem*, "Those films are not really about the future. I think both films were really about the present, which I tried to disguise with some future elements." Similarly, "People think I am a prophet and that Brazil described the world we're living in now a few years ago. But we were living in that world then; people just weren't paying attention the way they do now." Which is progress, of a sort.

"OH MY GOD, A MISTAKE!"

Sam discovers the Buttle/Tuttle error when he realizes the wrong bank account has been debited for the arrest (in this dystopia, you're charged money for being apprehended and detained by the state), and he visits Mrs. Buttle to give her a refund.

There, he spots her upstairs neighbor Jill, previously glimpsed only on Ministry surveillance camera footage. She resembles the woman of his dreams (literally). Jill has been trying to help Mrs. Buttle find out what happened to her husband, but her efforts have been obstructed by bureaucracy. What Jill doesn't know is that she's now considered a terrorist accomplice of Tuttle for trying to report Buttle's wrongful arrest. Sam approaches Jill, but she closes up, worried the government will track her down.

Then, Sam's heating unit has malfunctioned and he tries to get it fixed (it's those ducts again), but Central Services are characteristically uncooperative. Harry Tuttle (Robert De Niro), who's actually a 'guerrilla heating engineer' who used to work for Central Services but left because he disliked the paperwork, unexpectedly comes to his assistance. Unfortunately, Central Services workers arrive soon after. If Sam is found to have hired unofficial help, he'd be in trouble. But Sam manages to scare the workers away by asking them for the dreaded '27B/6' bureaucratic form. They flee (later, they return with the requisite form, discover Tuttle's handiwork, and effectively destroy Sam's apartment with their 'repairs').

The viciously totalitarian nature of the regime becomes ever more apparent as the film goes on, as Sam is drawn further into its endless, nightmarishly bureaucratic maze. But it's apparent from the beginning, in the frightening but funny scene in which black fatigued gas masked police storm the Buttles' apartment in the misleadingly named Shangri-la Towers, and bag and remove Mr. Buttle to 'help them with their enquiries' (Shangri-la of course is the mythical Himalayan utopia from James Hilton's 1933 novel *Lost Horizon*).

Gilliam is reportedly perplexed that *Brazil* is a favorite film on the American right. But it's not difficult to understand: it depicts a despotic state-run society, and it doesn't work. Worse than that, its manifold mistakes are magnified by the system's meanness and malevolence. Indeed, what starts Sam's journey into conflict with the regime is his attempt to behave humanely within it, by hand-delivering a refund check to Mrs. Buttle for her husband.

This is an oppressive but grossly inefficient central government, (dis-)organized into inhumanly huge, blame-shifting bureaucracies that claim the right to spy on, interfere in, and ultimately destroy, people's lives. Despite the government's claim on its propaganda posters ('Happiness: We're All In It Together'), just as in actually existing communist regimes we see there's still a favored, party-connected class (Sam's mother belongs to it). And clearly, the state doesn't tolerate individual initiative or entrepreneurship. Not at all.

Form 27B/6 is a reference to George Orwell, who lived in apartment 27B, Floor 6, in Canonbury Square in London, while writing parts of *Nineteen Eighty-Four*. In its setting and themes, *Brazil* obviously references Orwell's novel. Gilliam describes it as "the Nineteen Eighty-Four for 1984", despite claiming never to have read the book or seen any of its film adaptations (despite this, one of the working titles for *Brazil* was 1984 ½). *Brazil* also employs an Orwellian degradation of language; for example, 'Information

Retrieval' is the euphemistic name for the department that deals in interrogation and torture, along with the state's doublespeak slogans such as 'Suspicion Builds Confidence.'

Debate continues over whether Orwell was 'left or right,' and indeed the right can take some ideological comfort in Orwell's anti-Stalinism. But of course, Orwell fought for the Republicans *against* Franco's Nationalists in the Spanish Civil War, while also seeing firsthand how Soviet-backed communists suppressed revolutionary socialist and anarchist dissenters like himself. And in a letter to Francis A. Henson of the United Auto Workers union, Orwell made it clear that *Nineteen Eighty-Four* was intended "as a show-up of the perversions to which a centralized economy is liable, and which have already been partly realized in Communism and fascism", and moreover, that "totalitarianism, if not fought against, could triumph anywhere."

Gilliam himself, and *Brazil*, might be more difficult to characterize. He's no fan of 'political correctness,' 'health and safety culture,' and other favorite contemporary right-wing targets—and obviously, bureaucracy. At the same time, he's an instinctive anarchist. Like Orwell's novel, *Brazil* is anti-totalitarian. And also like Orwell, those on the left should be willing to acknowledge how authoritarianism can derive just as easily from socialist thinking as it can conservative ideology, including the utopian-dystopian idea that the bureaucratic central services can and should provide for our every want and need.

"COMPUTERS ARE MY FORTE"

Which brings us to so-called left accelerationism. In chapter 6, we discussed utopian technologism, Friedrich Hayek's cybernetics, and how the new right believed if we only allowed the market computer to work, there would be no limits. More recently there's been a resurgence of equivalent thinking on the left. As we'll suggest, this suffers from many of the same dangers and delusions as conservative cybernetics, especially in a context of collapse (this also applies to centrist technological solutionists, but let's focus on the left).

The term 'accelerationism' was coined (critically) by Benjamin Noys in 2010. It might have come from a 1967 sci-fi novel *Lord of Light* by Roger Zelazny, which includes a plot about a group of revolutionaries, the Accelerationists, who want to take their society to a higher level by transforming its attitude to technology (the novel is best-known today due to the 2012 film *Argo*; a screenplay based on it was used by the real CIA for the fake film production which provided the cover to rescue the hostages in Iran after the Revolution). Noys expanded on his critique in his *Malign Velocities:*

Accelerationism and Capitalism (2014), suggesting instead that we might all want to slow down a bit (or a lot).

Previously, we mentioned the Allende administration's Project Cybersyn, the attempt to introduce a cybernetically controlled economy for Chile cut short by the fascist-neoliberal coup, and how this echoed earlier interest in state planning using information technology. Despite the contemporary dominance of neoliberal thinking, such dreams haven't died.

In 1993, in response to neoliberal proclamations of the end of history, computer scientist W. Paul Cockshott and economics professor Allin Cottrell published *Towards a New Socialism*. This proposed a complex, planned socialist economy, inspired by cybernetics, Marx, and British operations research scientist Stafford Beer's 1973 model for Project Cybersyn (described in his 1972 book *The Brain of the Firm*). Computers would determine and manage the allocation of resources between socially owned enterprises, but there would still be a kind of market in consumer goods (we'll come back to consumerism later).

Twenty years later, Nick Srnicek and Alex Williams published their *#Accelerate: Manifesto for an Accelerationist Politics*, later expanded into a book, *Inventing the Future: Postcapitalism and a World Without Work* (2015).[2] Srnicek and Williams argue that automation makes progressive reforms such as a Universal Basic Income and dramatic reductions in working hours reductions not only feasible, but also necessary.

Against the idea of the end of history and the failure of a socialist future once thought inevitable, Srnicek and Williams aim to reclaim the future from the right, through a future-oriented left politics "at ease with a modernity of abstraction, complexity, globality, and technology." If full automation is inevitable, we shouldn't fear it, instead we should embrace it. We shouldn't restrain, but rather "unleash latent productive forces" (despite the title of their manifesto, Srnicek and Williams don't use the term 'accelerationism' in their book).

Socialism has always been a kind of utopian imaginary, sometimes a very popular one. One of the first utopian novels was Edward Bellamy's best-selling 1888 fantasy *Looking Backward*, about a socialist paradise in the year 2000 (we mentioned the backlash to Bellamy in chapter 6). More practically, as Adam Flynn has stated, "The great programs of the 20th century often began as fictional proposals, from moon landings to Social Security." The left has often owned the future, in seeking to both imagine and create it, and presenting its ideas as naturally futuristic, even historically inevitable. This is the tradition that Srnicek and Williams want to reclaim and reenergize.

Following Srnicek and Williams, there have been other leftists proposing similar ideas. Aaron Bastani's *Fully Automated Luxury Communism* (2019) proposes that automation could be the path to a postscarcity world of liberty, luxury, and happiness for everyone, since technological advances will reduce the cost of food, healthcare, and housing towards zero. Paul Mason, in *PostCapitalism: A Guide to Our Future* (2015), argues the digital revolution, including the ability to reproduce 'goods' such as software at virtually no cost, poses an existential threat to capitalism, reshaping our notions of work, production, and value, and destroying an economy based on markets and private ownership. Mason thinks this is already happening—in parallel currencies, cooperatives, self-managed online spaces, even Wikipedia—as examples of what the postcapitalist future could look like. Like others, he supports the idea of reducing work and paying everyone a basic income.

"GOOD LUCK, KEEP YOUR EYE ON THE BALL"

In chapter 2, we discussed how the futures promised by capitalism haven't materialized, not just in technology, but in work, leisure, and living standards. As Srnicek and Williams set out in their original manifesto, not only is capitalism unjust, but it is also a system that despite its propaganda actually often holds back progress. As Mark Fisher put it nicely, in contemporary capitalism, "A simulation of innovation and newness . . . cloaks [its] inertia and stasis."[3]

Accelerationism then represents an attractive strategy, theoretically and politically; it highlights neoliberal capitalism's stagnation and monopolism, and as discussed here, its faltering, failing futures. It outflanks capitalism's claims about progress, and confidently reasserts the left's lost futurism. This is why Fisher, among others, argued that accelerationism can function as an anti-capitalist strategy, and replace what he regarded as the contemporary left's tendencies towards "Canutism, its rhetoric of resistance and obstruction."

As Srnicek and Williams recognize however, accelerationism isn't new. The idea goes all the way back to Marx and Engels (Fisher argued that Marxism "is nothing if it is not accelerationist"). It's also influenced by 1990s cyberculture including cyberpunk, in looking forward to an advanced, disruptive technological future. But Srnicek and Williams deny they're techno-utopianists. They say they recognize technology on its own won't save us. Their point is that post-capitalism, and harnessing the full potential of technology, requires post-capitalist planning, that is, some central direction.

What they really bemoan is the collapse of previous popular imaginations of the left and being stuck in a backward-looking, defensive project, for example trying to rescue whatever's left of the mid-twentieth-century welfare state. Instead, they want to reassert the future, technological optimism, and the importance of utopian thinking. As they put it: "Rather than settling for marginal improvements in battery life and computing power, the left should mobilise dreams of decarbonising the economy, space travel, robot economies—all the traditional touchstones of science fiction—in order to prepare for a day beyond capitalism."

Obviously, we should harness technology in a much more equitable way to free people from hunger and other severe wants. And it's a powerful argument, for example made by Eddie Yuen, that environmentalists should focus on material abundance for the many as the way to defeat eco-fascist scarcity politics of the few.[4] As Theo Curtis similarly puts it, "To defeat the allure of the retrotopia offered by the right, the left must invent new utopias for people to believe in. Only a powerful and persuasive vision of a better future is capable of preventing a regression into a world of competing intolerant nationalisms."[5]

And yet, there's still a feeling of technologism, and technological determinism, here. Computerization and robotics are making ever more jobs redundant without inevitably producing new jobs to replace them. But despite arguing that the left should reclaim the future, in one sense Srnicek and Williams seem passively to accept that technology determines the future and there's little we can do about it. The only question is who it's used *for*.

So, accelerationism is not quite the radical break with industrial capitalism it claims to be. As Franco Berardi points out in *After the Future* (2011), the center-left and left's earlier futurism was really founded on capitalism's "myth of the future," in the experience of the continual expansion of knowledge and the economy—just that with collective planning and coordination, the future could be *even better*. But the most fundamental problem with accelerationism is how credible, even desirable, it is in a world in collapse.

"YOU DON'T EXIST ANYMORE. I'VE KILLED YOU."

Sam finds out Jill's records have been classified. The only way to access them is to be promoted to Information Retrieval. But Sam has previously turned down a promotion arranged by his ambitious, socially connected mother, Ida (Katherine Helmond), who's obsessed with the 'rejuvenating' plastic surgery of cosmetic surgeon Dr. Jaffe.

Around the same time, in Sam's dreams he battles a huge samurai made up of computer parts. After he defeats it, he removes its mask and sees his own face. Sam is battling the machine, the bureaucracy, in which he's complicit, but he dreams of escape (more darkly, Sam's dreams start to include victims of the regime, such as Mrs. Buttle).

As Sam tries to engage with his mother to ask again about the promotion, we're taken through Jaffe's surgery and onto a French restaurant where Mrs. Lowry lunches. Straight out of Orwell, we've already noted the persistence of an upper class in this supposedly much more rationally organized society; later, after an apparent terrorist bombing of the restaurant, privacy screens are placed around the snooty diners so that the dead and dying don't put them off their 'food.' But one way that *Brazil* certainly differs from *Nineteen Eighty-Four* is that it also satirizes consumerism—not something that was an issue when Orwell was writing of course.

It started with the ridiculous advert for ducts (does anyone really need them in "colors to suit your individual tastes"?) Sam's mother and the other ladies who lunch wear absurd fashions (Mrs. Lowry touts a leopard skin shoe on her head). We see a person in the street marching with a banner that reads 'Consumers for Christ.' In one of Sam's dreams, Santa Claus asks a little girl, "What would you like for Christmas?" She replies, "My own credit card." When another terrorist bomb goes off in a department store, passers-by don't even pause their shopping. Most darkly, there's even a scene towards the end of the film where an interrogator urges his victim to confess quickly, lest it ruin his credit rating. Later, Sam and Jill drive along a road away from some oil fields. The highway is lined with billboards packed together so that from the road it's not possible to see the devastated industrial landscape on the other side. Even as society has cratered around them, democracy has died, and the environment has been poisoned, people carry on consuming.

Which might make us question why left accelerationists seem to want to mimic a rampantly wasteful, unsustainable, capitalist consumer society—not just mimic it, but *multiply* it. Srnicek and Williams rightly argue that a reduction in working hours using automation is an important part of responding to climate change (unnecessary work and commuting are major contributors to carbon emissions, as well as being bad for us). They argue that post-work undercuts the primary reason for requiring growth and jobs (as traditionally understood), namely the link between income and work. But at the same time, accelerationists want to *increase* production by unleashing it from its 'capitalist fetters.' Left accelerationism then seems to rest on the same assumption as mainstream economics, that ever-expanding growth is necessary to increase social welfare.

Srnicek and Williams acknowledge they don't engage anywhere near enough with the environment, which is strange, because their original manifesto begins so powerfully by recognizing crisis and collapse in terms remarkably similar to *The Limits to Growth*: "At the beginning of the second decade of the Twenty-First Century, global civilization faces a new breed of cataclysm. . . . Terminal resource depletion, especially in water and energy reserves, offers the prospect of mass starvation, collapsing economic paradigms, and new hot and cold wars."

Srnicek and Williams suggest that whether full automation is compatible with ecological sustainability depends on replacing fossil fuels with renewable energy, the substitution of dwindling resources, the revision of wasteful manufacturing processes, and the elimination of exploitative labor practices. Which sounds true, but only *sounds* it, as we'll see.

For the moment, let's note the fundamental problem: increasing our productive capacity runs counter to the challenge set out by *The Limits to Growth* that we urgently need to reduce our use of natural resources. Left accelerationists seem to come from the opposite perspective, more like the cornucopians we discussed in chapter 5. There are no limits, as long as we program the (much more eco-friendly) machine to achieve what we want.

Indeed, the whole point of accelerationism is that we can achieve unprecedented power over presumed limits, including nature. As Srnicek and Williams state in *Inventing the Future*, "If complexity presently outstrips humanity's capacities to think and control, there are two options: one is to reduce complexity down to a human scale; the other is to expand humanity's capacities. We endorse the latter position." Bastani proclaims that, "Our ambitions must be Promethean because our technology is already making us gods—so we might as well get good at it."[6] Srnicek and Williams similarly call for a "Promethean politics of maximal mastery over society and its environment."

Of course, we could instead recognize we don't control nature, and we never have. But this reflects a longer-standing ambivalence in some leftist and Marxist thought about the environment. Until more recently, when the ecological crisis has become ever more apparent, parts of the left have been rather skeptical environmentalists. This included reactions to reports such as *The Limits to Growth*. At the time, some Marxists argued that environmentalism was primarily an elitist preoccupation of (Western) middle classes and was 'anti-poor,' even racist; others, in common with neoclassical economists, dismissed the Malthusian assumptions supposedly underlying such analyses.

Even if attitudes have shifted as the ecological crisis comes to be regarded as a consequence of industrial capitalism's rampant extraction, and there's been a reinterpretation of Marx's thinking that emphasizes his recognition of the tensions between nature, technology, and economic development, there's still a sense that 'the environment' is something to be fixed 'after the revolution', once the 'right people' in the state (rather than capitalists) take control of the levers (or keyboards). We'll come back to that.

"THAT TECHNIQUE . . . I'VE TRIED IT. A NICE EFFECT, BUT HIGHLY UNSTABLE."

As Ville Kellokumpu identifies, given our current predicament(s), accelerationism's vision of godlike technological capabilities for humanity seems "out of place" (to say the least). Instead, any visionary social project has to confront, in climate change and beyond, a material crisis of how we consume and relate to nature.[7]

And it's not, as some leftists seem to assume, that the crises brought on by ignoring limits will naturally hasten the end of capitalism and a new ecologically sustainable era. Firstly, as Saral Sarkar warns: "This double crisis [of climate change and resource limits] (I call it the pincer grip crisis) is not just the crisis of capitalism, as how most leftists would like to see it. In the long run, it will also inevitably cause the end of industrial society per se. This will happen whether we like it or not. Even a truly socialist industrial society would perish because of this crisis."[8]

Secondly, the left might better recognize the depth of the economic and environmental crisis overtaking industrial capitalism, but as we discussed in the previous chapter, the right understands the *political* dynamics of the crisis much better and how to exploit them. As the activist and educator DeRay Mckesson says, challenging leftist naiveté, "[T]he only thing that happens in a 'productive apocalypse' is an apocalypse."

Anyway, in terms of resource use, despite repeated warnings from *The Limits to Growth* onwards, rates of consumption, pollution, and despoliation are not declining or even holding steady, they are (of course) accelerating, as they have been since around 1950. This locks us ever more tightly into what has long been predicted. If it's still possible, which it probably isn't, climate stabilization alone would require a dramatic and sustained deceleration (called 'descent' in new economics circles) in Western (-style) levels of consumption.

And the thing about growth is, it's compound. Three percent annual growth in GDP sounds reasonable—it's commonly taken as the threshold for a healthily growing economy. What this means, though, is that the economy doubles in size every fifteen years. If resource use increases at the same rate, every fifteen years we've used as much resources as we have in *all of the preceding years put together*. To put it another way, once we've used up half of any resource, at the same rate of growth we're only fifteen years away from having used *all of it* (you can see why the cornucopians were so keen to claim that we'd just endlessly discover new resources). This is why physicist Albert Allen Bartlett said, "The greatest shortcoming of the human race is our inability to understand the exponential function."

It's not just collapse in environmental systems. It's conflict over natural resources, failed food production systems, and forced migration—it's collapse in our *civilizational systems*, which are more dependent on the environment than we typically recognize. Despite increasing public interest in collapse, we're still generally in denial about this, even in supposedly enlightened circles. There *is* a conspiracy of sorts among scientists and environmentalists, but it's not the one the right claims, rather it's a conspiracy against recognizing collapse, and dystopia. It's a collective understanding not to talk about it.

Some campaigners warn about spreading resignation if we state the truth. But this leads to a disjuncture for many people: can it *really* be so bad if experts such as climate scientists are just carrying on with their lives as normal? Jem Bendell even suggests it's difficult to get published in academic journals if you suggest collapse is inevitable, because it makes the reformist approach to 'sustainable development' redundant. As environmental engineer Josh Kearns states, "The widespread (over-) use of the term 'sustainability' reflects the subconscious desire of the affluent to maintain consumptive lifestyles and high socioeconomic status despite energy descent, ecological overshoot, financial breakdown, and economic contraction."

The irony, going back to the right's increasing conspiracism about the 'threat' from the left we discussed in previous chapters, is that elite conservatives don't really believe we're on the road to a leftist dystopia, but they use the idea to scare people into rejecting sensible policies that might enable us to at least reduce the extent of collapse. Meanwhile, the elite center-left acknowledges crises such as climate change, but then effectively deny these crises are likely to lead to a conservative dystopia, including by only proposing moderate policies (and resisting more radical reforms) as if these are sufficient to avoid collapse.

Indeed, as 'ecomodernist' and critic of the environmental movement Ted Nordhaus has pointed out, there's a certain emptiness to some center-left and left talk about a 'climate apocalypse.' If this really is an emergency, programs such as the Green New Deal are really rather modest—about public funding for renewable energies (including corporate subsidies), and regulations, and not for example the nationalization of energy sectors.[9] It also claims to reject technocracy while essentially being a technofix itself.

Truly recognizing collapse challenges both left accelerationism and this kind of liberalism. It says, technology and technocratic solutions won't be our way out of the maze.

"SOMEONE SOMEWHERE IS TRYING TO MAKE US CARRY THE CAN"

In Sam's dreams, he's a silver-blue Icarus-like figure, soaring above clouds. In Greek myth, Icarus was the son of master craftsman Daedalus, the creator of the Labyrinth. Icarus and Daedalus attempt to escape from Crete using wings that Daedalus makes from feathers and wax. Daedalus warns Icarus of complacency and then hubris, instructing him to fly neither too low nor too high. We know how it ends, of course. Icarus ignores Daedalus. The sun melts the wax in Icarus's wings. He tumbles out of the sky into the sea and drowns.

Complacency and hubris sum up our technological solutionism, or saviorism. Left accelerationists typically dismiss small-scale, localist environmentalism as backwards-looking "folk politics" (we'll return to these criticisms in the next chapter). But not only will an accelerationist society, even an accelerationist planet, still have to deal with cascading crises, it's likely to exacerbate them. Despite its apparent radicalism, left accelerationism is a form of business-as-usual, that we can keep producing and consuming, indeed it promises even more people can consume like middle-class (even wealthy) Western people, while a green technological revolution saves us from the consequences. Really then, accelerationism has the same problem as 'green capitalism,' in that it fails to consider the more fundamental system change that's necessary to avert (or now, reduce the extent of) collapse. The only thing that's certain to accelerate are crises. We're about to begin our own very steep, and very scary, descent.

But what about a 'great transition' to a much more sustainable society? Whether it's led by an enlightened state or not, we underestimate not only the scale and complexity of transition, but also how it would use enormous amounts of resources. Currently, fossil fuels provide more than three-quarters of global energy. The International Energy Agency (IEA) estimates it will cost $70 trillion to decarbonize the world's energy systems. We'd need a huge construction enterprise to (re-)power the planet: building one terawatt (one trillion watts) of solar energy power plants every year, each lasting for twenty-five years, would entail manufacturing, putting up, taking down, and either recycling or disposing an area of solar panels eight times the size of Los Angeles *every year*.

But we also neglect how many other systems underpinning contemporary civilization depend on fossil fuels. For example, oil is used for tilling fields, making herbicides and pesticides, and transporting crops to market. Current production levels are only possible due to synthetic fertilizers and a fossil fuel powered agroindustry.

Similarly, transportation depends almost exclusively on oil. From a transition perspective, as self-described 'energy skeptic' Alice Friedemann emphasizes, this is the liquid fuel heavy-duty transportation crisis. There are trillions of dollars' worth of trucks, ships, locomotives, and equipment essential for planting, harvesting, and transporting food, building and maintaining infrastructure, and mining. They all have diesel engines, which can last forty years and go a million miles, and we don't yet know how to replace all of them with green alternatives. As Friedemann puts it, civilization ends when transportation stops.

Then there's the resource costs of such a transition. As we discussed in chapter 4, *The Limits to Growth*'s problematique points to the confounding maze we're now in, where more industrial output might mean more money to spend on agriculture and healthcare, but also more pollution, which could damage health and food production. The same goes for a green transition. Water, iron ore, minerals, and fossil fuels—any transition would require enormous amounts of these resources. Solar panels, wind turbines, electric vehicles, and batteries, require metals that must be mined, refined using toxic processes, manufactured using further processes, and deployed.

It's controversial in green circles, but Ozzie Zehner's *Green Illusions: The Dirty Secrets of Clean Energy and the Future of Environmentalism* (2012), discusses various approaches to 'clean energy' and why they don't allow us any easy escape. That's because, as Zehner states, "We don't have an energy crisis. We have a consumption crisis."

So, back to *Brazil*. As David Sterritt puts it: "At the core of the film's social critique is Gilliam's insistence that contemporary culture is enslaved by its technologies, which waste more time and resources than they save. The government in Brazil makes things even worse, aiming at total control but achieving zero due to stupidity and incompetence at every level."[10]

Is this a regime we'd trust to resolve the crises through big technological fixes, when industrial-scale technologism has caused the problems in the first place? It's one thing to recognize we now live in Anthropocene, the geological epoch marked by significant human impact on the Earth's geology and ecosystems, and how this effectively makes us responsible for what happens to these systems. It's another to assume this means we can manage or direct, let alone 'fix,' these systems through even larger-scale technologies.

As Elizabeth Kolbert suggests in *Under a White Sky: The Nature of the Future* (2021), well-intentioned technological fixes risk making matters worse (Kolbert's title refers to one of the consequences of trying to reflect more sunlight: our blue sky could turn pale white). From a technological solutionist's point of view, the answer to one form of human intervention in the natural world (climate change) is to try to counteract it with another (trying to scrub carbon dioxide out of the atmosphere). As Kolbert asks, tongue-in-cheek, "What could possibly go wrong?" Of course, to avert even greater catastrophe, we might have to try some of these approaches. But to focus on them, in the hope they will allow us to carry on producing and consuming, is a delusion, and now an extremely dangerous one.

Left accelerationism has an inherent bias towards technofixes rather than the social-ecological transformation we need towards lower consumption and much lower carbon emissions. Bastani even advocates asteroid mining to discover new mineral resources ("the limits of the earth won't matter anymore—because we'll mine the sky instead") – a sci-fi "space fix," as Ville Kellokumpu criticizes it, that echoes figures such as Elon Musk.

Indeed, in one sense accelerationism is one big techno-fix, not just for environmental crises, but political, social, and cultural crises as well. Whereas studies such as *The Limits to Growth* suggest that growth is the problem, accelerationists argue that more growth is the solution. But we've already noted how a great transition to renewable energy, transportation, and food systems, among others, will require enormous resource use. Grand techno-fixes are likely to require even more. We're lost in the labyrinth once again.

"DOESN'T IT BOTHER YOU, THE SORT OF THINGS YOU DO AT INFORMATION RETRIEVAL?"

Trying to get access to Jill's file, Sam gets the job at Information Retrieval. There, he tries to use his colleague's computer to find the information. Sam also discovers his old friend Jack (Michael Palin) is an interrogator/torturer for the state, and he was the one who killed Mr. Buttle. Of course, characteristic of the ethical death that happens under totalitarian regimes, Jack denies responsibility for his own actions: "Information Transit got to wrong man. I got the right man. The 'wrong man' was delivered to me as the 'right man.' I accepted him on good faith as the 'right man.' Was I wrong?"

Having obtained Jill's records, Sam tracks her down. They share a romantic night together. Sam falsifies her records to indicate her death, which will allow her to escape—but not in time. They're apprehended at gunpoint. Charged with treason for abusing his new position, Sam is restrained in a chair in a huge, empty room, to be tortured by, yes, Jack, who of course complains he really doesn't want to interrogate his friend but has 'no choice.' Sam is also told that Jill was killed while resisting arrest.

As we discussed with *V For Vendetta*, the more systems struggle with scarcity, the more likely they are to resort to authoritarianism: scapegoating others, characterizing them as 'enemies within' or even 'terrorists,' trying to exert ever more control through propaganda, surveillance, and detainment, and corrupting the truth about everything.

Brazil depicts a society struggling to maintain normality, such as shopping, in its desperation to avoid acknowledging crisis. This is why Central Services is advertising ducts in a variety of colors—it's trying to give the populace the idea they live in a society of fabulous consumer choice, all the while everything crumbles (or blows up) around them. But as we know from Sam's experience, in reality they're hardly responsive to their 'customers' – instead they end up literally terrorizing them. And if the well-connected elite are eating colored mush for food, what on earth is everyone else eating, if they're eating at all?

Who knows, automated luxury communism might work and the computer might perfectly meet our wants and needs. But it seems unlikely, and the effort to make it happen might end up consuming society, as well as what's left of natural resources. And if, at the moment, we're increasingly critical about the amount of data that big tech collect about us, why wouldn't we feel the same when the state's 'central services' does the same?

In *Brazil*, all this decay and denial has led inevitably to despotism, and the metastasizing, paranoid, oppressive bureaucracy that's trying to keep a lid on it all. We even learn at one point that the Ministry of Information alone consumes *7 percent* of gross national product—hence the need for 'Information Retrieval charges' to try to recoup its out-of-control costs. Really, the only thing left accelerationism offers is to build a better dystopia.

"DON'T SUSPECT A FRIEND—REPORT HIM!"

So reads one of the state's propaganda posters. We've been no fan of Hayek here. But he was at least right to encourage us to look for the seeds of authoritarianism in political ideologies (he just neglected to include his own in this, of course).

As far as we can see, *Brazil*'s Britain doesn't have a Big Brother all-powerful leader. The problem is that everyone, nearly everyone anyway, is compliant with the failing, crisis-prone system but is too scared to point out its mistakes, the regime brutally punishes anyone who does, and no-one in the bureaucracy takes responsibility for anything.

Brazil resembles what Guillermo O'Donnell termed 'bureaucratic authoritarianism,' his characterization of some regimes in South America between the 1960s and the 1980s, including Brazil. Unlike many dictatorships in the region centered on cults of personality around a single leader, Brazil experienced a growing set of complex and interlocking restrictions imposed by military rule and substantial repression, but also a bureaucratic, technocratic approach to policymaking. Whenever rules got in the way, it was common to work out a *jeito*—a way to get things done by working around the system.

A left accelerationist society would still struggle to deal with the cascading crises discussed here, whereas, as discussed in the previous chapter, a fascist regime would use the fear and scarcity they generate for its own ends. And a left accelerationist program would inevitably have to confront rightwing resistance and disruption, and most likely terrorism. Under such incredible pressures, would it remain free and democratic?

There's an unnerving power gap in left accelerationism, a desire to wish away 'politics' in the same way that Hayek and his fellow neoliberals did (or tried to, anyway). Don't worry, the machines will sort it out. Plenty will lead to peace. Except, of course, it won't. Even if the machines could, is an algorithmic socialism really what we really want? Because there *is* a seed of authoritarianism in left accelerationism. As Connor Owens says: "Full automation and common ownership of technologies won't be that liberating if control over those technologies remains in the hands of the state, most likely . . . directed [by] . . . a new 'techno-bureaucracy' . . . – the 'savants' Mikhail Bakunin warned of in his *God and the State*."[11]

"SORRY, I'M A BIT OF A STICKLER FOR PAPERWORK"

This is not to fall victim to the conservative caricature of bureaucracy as inevitably bad, let alone (as Hayek effectively did) inherently totalitarian. In some form, bureaucracies are needed in any functioning, and humane, society, and the right conveniently ignores the vast bureaucracies found in large corporations. But left accelerationism *does* pose a danger of constructing huge, centralized bureaucracies, with near-total information collecting powers, and perhaps worse.

Of course, as a dystopian satire, *Brazil* is an extreme depiction. But if we criticized the new right for the seeds of dystopia within its utopian cybernetic dream, it's only honest we do the same for the left. As some on the left promote 'crisis communism,' it's worth going back to Benjamin Noy's concerns about accelerationism. Noys noted that Lenin embraced the management theories of Frederick Taylor and dreamed of a world in which productivity gains would free workers to relax. During the 1930s, Stalin measured progress towards utopia through the rising output of products. A centrally planned economy would be much less wasteful, and much more productive, than capitalism. But such policies resulted not in universal leisure, but in the Kafkaesque machinations, surveillance, and repression of Soviet bureaucracy.

In the 1950s, after Stalin died, Communist reformers were able to consider how cybernetic thinking and information technology could be used to create a more responsive, decentralized, and democratic planning and coordination—a new Soviet future. As Richard Barbrook (we mentioned his description of the 'Californian Ideology' in chapter 6) describes in his book *Imaginary Futures* (2007), Khrushchev promised that within two decades the Russian people would be living in a postindustrial paradise of cybernetic communism. But ultimately this dream ran up against the limitations of technology, and more importantly, the Communist Party's commitment to hold onto its monopoly on power.

So, even if a left accelerationist future is technologically plausible, is it politically desirable? Despite attempts to envisage a more democratic cybernetic system, it's still fundamentally pro-technologism, pro-big planning, anti-localist, and potentially anti-liberty. In response to conservative cybernetics, the left shouldn't lose itself in an equivalent fantasy of a fully automated future. Instead, it should seek to create a more human, technology-using, but democratically determined one. As Joshua Mostafa suggests (reviewing Bastani's book): "The kind of epistemological modesty that will not only ward off the tyranny of the central committee, but also liberate us from the yoke of the market and the data centre, rejects both the hubris of omniscience and mystic reverence for 'spontaneous order'. . . . There are no easy short cuts to radical social and political change. The revolution will not be automated."[12]

"IF YOU HAVE ANY COMPLAINTS WHICH YOU'D LIKE TO MAKE, I'D BE MORE THAN HAPPY TO SEND YOU THE APPROPRIATE FORMS"

As Jack is about to start his interrogation, Tuttle and other members of the resistance break in, shoot Jack, rescue Sam, and blow up the Ministry building. Sam flees, but Tuttle is caught up in a tornado of paperwork from the Ministry.

In increasingly bizarre scenes, Sam stumbles into the funeral of his mother's friend, who has died following excessive cosmetic surgery. Sam discovers his mother now resembles Jill, and is too busy being fawned over by young men to care about her son. Police disrupt the funeral, and Sam falls into the open casket and through a black void.

Landing in a street, he's pursued by hideous baby-masked figures, some pushing shopping carts covered with coffins (deathly consumerism, again). He escapes by climbing a pile of ducts. His hand finds a doorknob. He steps inside the door and is surprised to find himself in a truck driven by Jill. They drive through the night. By morning, they're parked in a beautiful pastoral countryside. The camera tracks back and up, into the sky, the earth unrolling in all directions, music swelling triumphantly.

Yet again, our heroes have escaped the dystopian city to the utopian countryside. Except they haven't. In reality, Sam is still strapped to the chair. Realizing he's descended into seemingly blissful insanity, Jack and Mr. Helpmann declare Sam a lost cause and leave. Sam remains in the chair, smiling, quietly humming *Aquarela do Brasil* to himself.

Unfortunately, the studio didn't exactly love the ending. Nor did preview audiences. *Brazil* was produced by Embassy International Pictures and was released in Europe and internationally by 20th Century Fox, where it did well. But U.S. distribution was handled by Universal, who were hoping for a mainstream hit, and they wanted changes.

As with *Blade Runner*, a happy version of the ending was produced, this time against the director's wishes. The studio negotiated with Gilliam to do a slimmed-down cut, but he only made minor edits and refused to change the film any further. Infuriated, the president of MCA-Universal, Sid Sheinberg, delayed its release and insisted on a dramatic re-edit (Sheinberg was notorious for interfering; you'll recall from chapter 2 he urged Robert Zemeckis to change the title of *Back to the Future* to 'Space Man from Pluto'). This is the infamous 94 minute 'Love Conquers All' cut of *Brazil* (down from 142 minutes), where Sam gets the girl instead of descending into insanity.

After a lengthy delay with no sign of the film being released, Gilliam took out a full-page advert in *Variety*. It read: "Dear Sid Sheinberg, When are you going to release my film, 'Brazil'? Terry Gilliam." Gilliam also showed the film to West Coast film critics (in the backroom of a gun club, no less). The film won numerous Los Angeles Film Critics Association awards, forcing Universal to relent. But as we saw, many East Coast critics were sniffier.

"OF COURSE YOU WANT SOMETHING. YOU MUST HAVE HOPES, WISHES, DREAMS."

As with *12 Monkeys*, we can read the ending in different ways. In one sense, Sam beats the system. He lives fully in his dreams; they can't reach him now, and he'll never be made to love Big Brother. Apparently, this is Gilliam's interpretation too. By reaching a place where he can't be hounded anymore, Sam wins an ironic victory over the regime.

In this vein, in his review of *Brazil*, Salman Rushdie contrasts the creative power of dreams against a dark reality. Rushdie suggests that the film's moral is that: "[T]he world of the imagination is a place into which the long arm of the law is unable to reach. . . . This idea—the opposition of imagination to reality, which is also of course the opposition of art to politics—is of great importance, because it reminds us that we are not helpless; that to dream is to have power."

And yet, Sam's escape is only into fantasy. Sam was always a bit of a fantasist, but at the beginning of the story this was his respite, his way of coping with the system. When he saw a chance of the fantasy becoming real, or so he thought, he became aware of the distance between his dream and reality. As they say, it's the hope that kills you.

No-one really *likes* the downbeat ending, but it rings far truer than any of the alternatives. For a considerable time to come, a future at any scale in which sanity prevails, and we use technology to meet people's real needs much more equitably, is a delusion. 'Somewhere in the 20th century,' as *Brazil*'s title card has it, we created our current civilizational dilemma, and like Sam, we can't easily escape its nightmare maze. The dream of reaching 'Brazil,' a mythical land of freedom, joy, and leisure, will remain just that. We're much more likely to crash land in *Brazil*—a scarcity-ridden, paranoid, dysfunctional, all-to-real dystopia. As Jill remarks earlier on, there's nowhere to escape *to*.

Terry Gilliam is one of the great film fantasists, and *Brazil* is arguably his masterpiece. But the real message of the film seems to be that there's no escape, not inside the system, and especially not in fantasies. Only Tuttle, and then Jill, offer any resistance to the regime, unsuccessfully. But as Fred Glass suggests in his review of *Brazil*, Jill and Tuttle's failure "is that of the individual response to a collective problem, to a social structure capable of absorbing or neutralizing individual revolts."[13] And as Glass wryly notes, it's only during Sam's false flight from the torture chamber that we are finally able to read the slogan on a previously obscured billboard depicting a happy family driving in their car. It says: "We're all in this together."

NOTES

1. Salman Rushdie, "The Location of Brazil," 1985, reprinted in *Imaginary Homelands*, Salman Rushdie (London: Granta Books, 1991).

2. Alex Williams and Nick Srnicek, "#ACCELERATE MANIFESTO for an Accelerationist Politics," *Critical Legal Theory*, May 14, 2013. http://criticallegalthinking.com/2013/05/14/accelerate-manifesto-for-an-accelerationist-politics/ Also Nick Srnicek and Alex Williams, *Inventing the Future: Postcapitalism and a World Without Work* (London: Verso Books, 2015).

3. Mark Fisher, "Terminator vs. Avatar: Notes on Accelerationism," Mark Fisher blog, presented at the Accelerationism symposium, Goldsmiths, University of London, September 14, 2010. https://markfisherreblog.tumblr.com/post/32522465887/terminator-vs-avatar-notes-on-accelerationism.

Also Mark Fisher, "Terminator vs. Avatar," in *#Accelerate: The Accelerationist Reader*, Robin Mackay and Armen Avanessian (eds.) (Falmouth: Urbanomic Media Ltd, 2014), 335–46.

4. Eddie Yuen, "The Politics of Failure Have Failed: The Environmental Movement and Catastrophism," in *Catastrophism: The Apocalyptic Politics of Collapse and Rebirth* Sasha Lilley, David McNally, Eddie Yuen, and James Davis (Oakland, CA: Spectre/PM Press, 2012), chapter 1.

5. Theo Curtis, "Whatever Happened to the Future? Lessons from the 1930s," *In The Long Run*, July 18, 2018. www.inthelongrun.org/articles/article/whatever-happened-to-the-future-lessons-from-the-1930s/.

6. Aaron Bastani, *Fully Automated Luxury Communism* (London: Verso, 2019), 189.

7. Ville Kellokumpu, "Fully Automated Luxury Communism: A Manifesto By Aaron Bastani," *Society & Space*, September 3, 2019. www.societyandspace.org/articles/fully-automated-luxury-communism-a-manifesto-by-aaron-bastani.

8. Saral Sarkar, *The Crises of Capitalism: A Different Study of Political Economy* (Berkeley, CA: Counterpoint, 2012).

9. Ted Nordhaus, "The Empty Radicalism of the Climate Apocalypse," *Issues in Science and Technology* XXXV, no. 4, summer (2019). https://issues.org/the-empty-radicalism-of-the-climate-apocalypse/.

10. David Sterritt, "Brazil: A Great Place to Visit, Wouldn't Want to Live There," Essays for the Criterion Collection, *Criterion*, December 4, 2012. www.criterion.com/current/posts/2583-brazil-a-great-place-to-visit-wouldn-t-want-to-live-there.

11. Connor Owens, "Social Anarchist Futures," *Solarpunk Anarchist*, May 2, 2018. https://solarpunkanarchists.com/2018/05/02/social-anarchist-futures/.

12. Joshua Mostafa, "The Revolution Will Not Be Automated," *Sydney Review of Books*, July 23, 2019. https://sydneyreviewofbooks.com/review/zuboff-bastani/.

13. Fred Glass, "Brazil (1985) By Terry Gilliam," *Film Quarterly* 39, no. 4, summer (1986), 22–28. https://scrapsfromtheloft.com/2019/03/12/brazil-terry-gilliam-review-fred-glass/.

Chapter 9

"Come with Me If You Want to Live"

It's Too Late to Avoid Terminator 2*, But There's Still Time for a Human Resistance*

Terminator 2: Judgment Day (1991) might seem to suggest we can change the future, but the Terminators keep coming back from it, so maybe not. Equally, it's not really about the dangers of killer A.I., though obviously there is that. Rather, it's about accepting the reality of the future, and the loneliness that most other people can't or won't see it too. 'Accepting' collapse means building our own, much more secure social infrastructure. Radical decentralization and re-localization represent the only viable non-authoritarian alternative to disaster capitalism and emergency centralism, and the best, really the only, foundation for a new left politics of collapse. The future isn't over, not completely. Magical technologies won't save us, nor will charismatic leaders touting simplistic solutions. But creativity, communities, and commitment will help to save some of us, in some times and places. Just how many times and places is still up to us.

"THERE'S NOT MUCH TIME LEFT IN THE WORLD NOW, SARAH"

'Los Angeles, 2029 A.D.' Ominous, low-key music. The camera pulls up and over a graveyard of burnt-out cars, human skeletons still at the wheel, tracks through a charred children's playground, then across a sea of skulls. The camera comes to rest on one of the skulls. We hear the numbed voice of Sarah Connor (Linda Hamilton): "3 billion human lives ended on August

29th, 1997. The survivors of the nuclear fire called the war Judgment Day. They only lived to face a new nightmare: the war against the machines."

CRACK. A metallic 'foot' crushes one of the skulls as it marches forward. Pan up, it's a Terminator machine, scanning for new targets, shooting purple plasma fire from its gun. Other Terminators advance across the battlefield. A robot tank rolls relentlessly forward, while Hunter-Killer attack ships hover above. It's the Future War. The human resistance army is fighting bravely. They manage to destroy a few enemy vehicles, but they're suffering heavy losses. The machines' laser cannons are precise and unrelenting.

The human army flees in retreat. Cut to a resistance stronghold. We follow a man—we'll learn who he is in a moment—through an underground tunnel and up onto a vantage point, where he can survey the battle through binoculars. Sarah's voiceover continues: "Skynet, the computer that controlled the machines, sent two terminators back through time. Their mission: to destroy the leader of the human resistance. John Connor, my son."

The camera closes in on John's face, a battle-hardened warrior who might save humanity. Sarah concludes: "The first Terminator was programmed to strike at me in the year 1984, before John was born. It failed. The second was set to strike at John himself when he was just a child. As before, the resistance was able to send a lone warrior, a protector for John. It was just a question of which one of them would reach him first."

We fade to a wall of fire, the kind we imagine from a nuclear detonation, underscored by composer Brad Fiedel's pulsing synthesizer theme. We see the playground again, now ablaze, swing seats still in motion, as if the children have just been incinerated. Out of the flames emerges the terrifying silver skull of a Terminator, its glowing red eyes staring us down. And then the title: *Terminator 2: Judgment Day*.

It's a great opening sequence. But it's the few seconds of scenes before this that are the most haunting: brief glimpses of *before* the end of the world, when people were carrying on with their lives, oblivious. Automobiles streaming down a massive Los Angeles freeway, people crossing a busy street, children on swings against the smoggy city skyline. No-one with any idea that Armageddon is just moments away (well, one person knows, of course). A few of these former city dwellers now fight against the machine army. Most of them didn't survive.

It's just contemporary industrial capitalist civilization: urbanization, transportation, heaving populations, everyday pollution. This Los Angeles might be overcrowded and ugly, but it's teeming with life, home to so many people. And then the scenes slow, bleached out by bright white light: the nuclear detonations of Judgment Day. And just like that, the world we knew, the world we assumed, is over.

At the opening of *Terminator 2*, Sarah speaks with certainty, even though she hasn't actually seen the future herself, it only exists in her imagination. Not that it isn't real; in the previous film, she was hunted by a machine from another time, so she knows it's coming.

As we'll see, Sarah acts as if the future is unavoidable. All she can do is prepare for it, and this knowledge takes a heavy toll. But then the possibility opens up that the future might be changed. Through dystopian stories, we've also glimpsed the dark future. Some of us might feel haunted by it. But there's the possibility here too, not to protect a past that's over, but to create a different future—a human one.

"NOW DON'T TAKE THIS THE WRONG WAY, BUT YOU ARE A TERMINATOR, RIGHT?"

A truck stop, the middle of the night. The breeze scatters trash. Electricity sparks around the parking lot. A bright orb appears. A flash of light, then we see a naked figure crouching. The Terminator (Arnold Schwarzenegger) rises to its feet. We see through its red tinted point-of-view, the machine scanning the environment as it approaches a biker bar. In a menacing but funny scene, the Terminator 'borrows' a leather biker outfit from one of the patrons. And a Harley-Davidson. Oh, and he'll take a pair of sunglasses too.

Somewhere in the city, in another parking lot, a police officer checks out a strange electrical disturbance. As he inspects a hole burned through a chain link fence, he's grabbed and knocked out by a naked man (Robert Patrick). The man takes the cop's clothes and gun, and uses the police car's computer to find the address of one John Connor. Apparently, John has quite the track record of juvenile delinquency.

These figures arrive just like the first film, *The Terminator* (1984), itself: out-of-nowhere, and dramatically. In Los Angeles, 1984, a cyborg assassin, the Terminator (a T-800, also called a Cyberdyne Systems Model 101), arrives from 2029. It's a killing machine, a metal endoskeleton covered by a layer of living tissue that makes it appear human.

Terminator 2 riffs on the opening of its predecessor: two figures sent back in time, one to kill, one to protect the future leader of the human resistance. But at the start of the first film, John has yet to be conceived, and his mother is blissffully unaware of his, and her, signifiance. She's told about it, and the dystopia to come, by her human protector, future solider Kyle Reese (Michael Biehn). Understandably, she struggles to accept it.

The Terminator was a small independent film with a punk attitude made under somewhat guerrilla filmmaking conditions. It was conceived as a slasher horror movie, and it's almost all set at night. As the opening caption puts it, "The machines rose from the ashes of the nuclear fire. Their war to exterminate mankind had raged for decades. But the final battle would not be fought in the future. It would be fought here, in our present . . . tonight." Even converted into a sci-fi story (*Mad Max 2* was one influence), it retains a nightmare quality, a palpable feeling of being hunted by an unknowable, unstoppable monster. Appropriately, the image of the Terminator came to James Cameron, the director of the first two *Terminator* films, in a fever dream.

After seeing *Star Wars* (*A New Hope*) in 1977, the Canadian-born Cameron quit his job as a truck driver (there are lots of trucks in his *Terminator* films), and worked his way into the movie business. Technically minded and notoriously demanding, he made a science fiction short, *Xenogenesis* (1978), then became a model maker for low-budget producer Roger Corman. He later worked on films such as *Battle Beyond the Stars* (1980) and *Escape from New York* (1981), eventually getting directing work on 1981's *Piranha II: The Spawning* (Cameron was fired for his perfectionism, but continued to help out on the film).

Three years later came *The Terminator*. It topped the American box office for two weeks, established Cameron's directing career, and confirmed Schwarzenegger's status as a leading man. Sci-fi writer Harlan Ellison loved the movie but believed it was based on one of his short stories and an episode he'd written of the TV series *The Outer Limits*. He threatened to sue. The similarities between the stories seem rather limited, but Orion settled and gave Ellison an acknowledgment in later prints of the film (Cameron reportedly disagreed with the decision). Nevertheless, an iconic sci-fi series was born.

"I'LL BE BACK"

Both *The Terminator* and its sequel have the brilliantly simple structure of a chase made up of a series of dramatic action set-pieces, and it's the humans being chased. But what really elevates the films are *why* it matters—nothing less than human extermination is at stake—and the psychological and emotional impact on Sarah, Kyle (a man literally out of his own time), and in the sequel, Sarah's son John as well.

In the first film, Kyle explains to Sarah that in the near future an A.I. defense network named Skynet, created by a company called Cyberdyne Systems, becomes self-aware and initiates a nuclear holocaust (yet again, a computer succumbs to paranoia). Future John rallies the survivors and leads the resistance against Skynet and its machine army. When Sarah asks if Kyle can destroy the cyborg, he says he doesn't know. It seems hopeless.

In the space of a few hours, Sarah, a kind, seemingly happy-go-lucky diner waitress, is forced to come to terms with a terrifying new future. Not only that, but she's also told she has a crucial role in history—a history that hasn't happened yet. As Kyle describes it to her: "Most of us were rounded up, put into camps for orderly disposal. . . . We were that close to going out forever. But there was one man who taught us to fight, to storm the wire of the camps, to smash those metal motherfuckers into junk. He turned it around. He brought us back from the brink. His name is Connor. John Connor. Your son, Sarah . . . your unborn son."

Understandably confused and angry, Sarah replies, "Oh, come on. Do I look like the mother of the future?" But Kyle explains that John, who sent him back into the past, gave him a message for her: "Thank you, Sarah, for your courage through the dark years. I can't help you with what you must soon face, except to say that the future is not set. You must be stronger than you imagine you can be. You must survive, or I will never exist."

Kyle and Sarah defeat the Terminator, but Kyle dies. Months later, a pregnant Sarah is traveling through Mexico. At a gas station, a boy takes a Polaroid of her, the same photograph John will give to Kyle in the future. The boy's father says there's a storm coming in. Sarah says, "I know." The point isn't just the implication of battles to come, it's that Sarah acknowledges the new future. But of course, acknowledging it is just the start.

Back to *Terminator 2*. The teenage John's foster parents are struggling with him. Ignoring their demands that he clean his room, he takes off on his motorbike to the mall. Meanwhile, we find out what's happened to his mother. Sarah is detained in a mental hospital. She's doing pull-ups on her bed frame, as psychiatrist Dr. Peter Silberman (Earl Boen) – who also interviewed Kyle in the first film after he's arrested by the police—shows interns around the hospital. Silberman patronizingly describes Sarah's "intricate delusion" about robot executioners and a future apocalypse. The orderlies force Sarah to take her meds.

The man we saw the previous evening, now wearing the cop's clothes, turns up at John's home and asks his foster parents for his whereabouts. They assume John's in trouble again. They also mention that a "big guy on a bike" was there earlier asking about him. The man tells them not to worry about the other guy.

John hacks into an ATM using his laptop. With his friend he takes off for the mall, avoiding the Terminator by a few blocks. During a review session later that day, Sarah tries to pretend she no longer believes her 'delusion,' but Silberman sees through her deception, denies her transfer to a minimum-security ward (she's tried to escape before), and even visitation rights (she just wants to see John). Angry and helpless, Sarah attacks Silberman, before being restrained and drugged again.

Terminator 2 was scripted by Cameron and William Wisher Jr., Cameron's childhood friend and writing partner (Wisher had contributed to *The*

Terminator and penned its novelization). It might be a blockbuster sci-fi action thriller with groundbreaking special effects, but it has an unusually convincing psychological-emotional core—and a message.

Reviewers largely agreed. Janet Maslin in *The New York Times* thought that: "In the end, it amounts to quite the pistol-packing plea for peace." Other reviewers were less convinced, for example Richard Corliss in *Time* magazine thought that Cameron couldn't deliver on the promise of his premise, despite its breathless beginning and serious scope.

Terminator 2 is certainly big. It's rumored to have cost at least $94 million, at the time the most expensive film ever made (in unadjusted dollars). It was the highest-grossing movie of 1991 and is now regarded as a multiplex classic. But it's also one of American Film Institute's Top Ten Sci-Fi films of all time, and as Richard Roeper in the *Chicago Sun-Times* reflects thirty years later, the film remains relevant, perhaps more than ever: "Doomsday scenarios are still conceivable, technological advancement is still outstripping human morals and capabilities, and PTSD and our place in the world are evergreen concerns in an escalating Covid-19 pandemic. The authorities prevaricate and throw us conflicting messages. There is still no fate it seems but what we make ourselves."

"WAIT A MINUTE HERE. YOU'RE TELLING ME THAT THIS THING CAN IMITATE ANYTHING IT TOUCHES?"

John, his friend—and the two figures searching for John—converge on the mall. In a videogame arcade, John plays *Missile Command*, the 1980 classic (originally titled Armageddon) in which players desperately try to defend their bases against incoming attack. Inevitable defeat leads to a screen starkly declaring "THE END." The game's creator, Dave Theurer, based it on his persistent nightmares about nuclear attack; as we'll see, Cameron (and many of us who were old enough at the time) shared similar fears.

The 'cop' finds the arcade. John's friend pretends he doesn't know him, but tips John off. John runs out the back door into a hallway. Suddenly, he's caught between the Terminator and the cop, both advancing towards him from opposite ends of the corridor.

The twist, revealed in advance marketing (Cameron presumably thought it would leak anyway), is that the Terminator is there to protect John, and the 'cop' is a new, even more advanced machine, a T-1000, programmed to assassinate him. A terrific, thumping fight ensues between the Terminator and the T-1000, in which we see just how indestructible this new foe is: shotgun rounds only dent, and barely delay, this upgraded metal monster.

John gets to his bike. A chase ensues. The T-1000 has hijacked a truck and bears down on John, while the Terminator tries to catch up on the Harley. The Terminator manages to blow up the T-1000's truck, and John and the Terminator escape, but only for the moment. The T-1000 emerges from the wreckage and morphs back into his cop form.

The most obvious way to interpret the *Terminator* series is as a warning about killer robots, and a necessary one. Warfare is being automated, and robots are creeping into policing. Big tech is working, often secretly, with states on defense and intelligence systems. And there's the pervasive surveillance we've discussed in previous chapters.

But as Maxim Pozdorovkin, director of the documentary *The Truth About Killer Robots*, has suggested, says, the Terminator trope is unhelpful—it directs us to think about a future "one day" in which malevolent A.I. might hurt us, rather than the much more pervasive effects of say automation today on jobs, skills, and human dignity. As Cameron says about *Terminator 2*, "What the film is about, on the symbolic level, is the dehumanization we do on a daily basis." It's about technology done to us, rather than under our control—through automation, algorithms, surveillance, disinformation, and manipulation. Which means it's something that *is* actually under human control; it's to do with corporate and state power, a lack of regulation, and democratic deliberation, in other words, technology in a neoliberal capitalist context in which ordinary people feel powerless to shape the future (as we noted in chapter 6).

Elites want us to believe that 'technology,' by which they mean *their* view about how technology should be developed and deployed, can't be questioned, it just *is*. To put it in the words of Kyle in the first film, industrial capitalism "can't be bargained with, it can't be reasoned with, it doesn't feel pity or remorse or fear, and it absolutely will not stop . . . ever, until you are dead." And if the T-800 is industrial capitalism, the T-1000 is accelerationist neoliberal capitalist technologism: even more destructive, deceptive, shape-shifting, and even more unstoppable.

We've already discussed how the new right rose up in reaction to new left environmentalism. At the risk of making them sound cooler than they are, the new right are like Terminators sent back from a dystopian future to kill off any resistance in the past, our present—to prevent us from changing the future by making even *thinking* about alternative futures, let alone creating them, more difficult. They've made judgment day inevitable.

Without many of us realizing it, we've been in a fifty-year battle with the machines—the nightmare of market-machine thinking—and the human resistance has been losing.

"DREAMS OF CATACLYSM, THE END OF THE WORLD ARE VERY COMMON"

After their escape from the T-1000, John tells the Terminator to pull over. He has some questions. The Terminator confirms what it is, and informs John it was sent by his future self to protect him in the present. The Terminator also explains their foe is an advanced prototype made of 'mimetic polyalloy,' liquid metal, that can imitate anything of equal size it comes into contact with (but not, ironically, complex machines). John wants to stop at his house, but the Terminator assures him the T-1000 will try to "reacquire" him there. The Terminator advises they get out of the city and avoid the authorities. Nevertheless, John insists on calling his foster parents to warn them. His foster mother answers, but it's the T-1000 impersonating her. The Terminator tells John his foster parents are dead.

Meanwhile, at the mental hospital, a police detective shows Sarah some pictures taken earlier that day: it's the Terminator at the mall. The police tell her John is missing and his foster parents have been murdered. Sarah doesn't respond. Silberman tells them she's lost touch with reality, but actually she's plotting her escape. She has to save John.

Of course, John now knows that everything his mother told him was true. He wants to rescue her. The Terminator refuses, explaining the T-1000 will likely copy Sarah and wait for John to come to her. But John insists, realizing the Terminator has been programmed to obey him. They head to the hospital. As the Terminator anticipates, the T-1000 is already there, looking for Sarah. She's taken Silberman hostage and forces the guards to open the exit. John and the Terminator arrive (John has instructed it not to kill anyone). Sarah makes it to the elevators, but to her horror, her nightmare appears: the Terminator. Then she sees John, who explains that this Terminator is here to protect him. John, Sarah, and the Terminator flee from the T-1000, only just escaping by accelerating away in a car. The three of them hide at an abandoned gas station.

Sarah reprimands John for risking his life trying to save her; he's "too important" for that. John is hurt by her cold reaction. Sarah is traumatized, unable to connect with her son, fixated on the need for him to survive to save a human future, but with her own humanity hurt. She even agrees with the Terminator that her life is not a "priority." Her trauma stems not just from her terrifying encounter with the first Terminator in 1984. No, she's also suffering from apocalypse PTSD, from a mass extermination that hasn't happened yet, but she knows lies down the road.

Kyle once said to Sarah that "Pain can be controlled, you just disconnect it." That only works for so long, though. Sarah's torment is a dramatic depiction of something that more people are feeling in our world. We discussed the increasing public recognition of collapse in chapter 7. But it's still crazy making, and lonely, to carry the knowledge that civilization as we know it is ending, in a world seemingly oblivious to what's coming.

Perhaps it's true that, as Derrick Jensen, Aric McBay, and Lierre Keith put it in *Deep Green Resistance* (2011), "[W]e know in our bones, if not in our heads, that this culture will not undergo any sort of voluntary transformation to a sane and sustainable way of living."[1] Maybe, though, we add to our feelings of anxiety and terror by continuing to hang onto a present that in reality has already gone. As Natalie Osborne writes, in the context of environmental destruction: "Perhaps release may come from considering that we are not in the midst of a desperate, heated, chaotic battle to save The World. In fact, we lost that battle, lost that future, lost that world. We. Have. Lost. Take a breath. Now what?"[2] To Osborne, and others, the first imperative is to grieve. But grieving isn't the end, or resignation. It's the start of acting according to a new reality.

Then there's John. As Cameron asks, "What does Jesus think when he's ten years old and you tell him he's the son of God. Doesn't that mess you up? Doesn't that mess up your mother?" John's innocence has been taken away from him before the film starts: "We spent a lot of time in Nicaragua and places like that. For a while there, [Sarah] was with this crazy ex-Green Beret guy, running guns. Then there were some other guys. She'd shack up with anybody she could learn from so she could teach me how to be this 'great military leader.' Then she gets busted. . . . She tried to blow up a computer factory, but she got shot and arrested."

We imagine years of chaotic living. John's been hurt, and presumably embarrassed, by his traumatized, imprisoned mother. *Of course* he's been acting out. Then he finds out she was right all along: the world really is ending and he's the future leader of the human resistance. But only if he can destroy the seemingly invincible machine that's relentlessly hunting him.

All in all, he takes it pretty well. John can still connect. He even bonds with the Terminator. He's still alive, still human, and now he's going to start fighting for the future.

"EVERYBODY DIES. YOU KNOW I BELIEVE IT, SO DON'T FUCK WITH ME!"

At a gas stop, John sees some children playing with toy guns. He ponders that humans probably won't make it. The Terminator comments that it's in our nature to destroy ourselves.

The Terminator explains that a cybernetics expert at Cyberdyne Systems corporation, Director of Special Projects Dr. Miles Dyson (Joe Morton), invents a microprocessor that will revolutionize computing. The Skynet program goes online on 4 August 1997. It learns at an exponential rate (the accelerationists' singularity idea we noted in chapter 6), and becomes self-aware on 29 August, prompting humans to try to shut it down.

The shadow of militarism and nuclear war runs through Cameron's films. In *The Abyss* (1989), military paranoia over a stranded American nuclear submarine nearly ends in a devastating confrontation, until a surprising intervention. In *Aliens* (1986), the ravaged survivors plan to exterminate the xenomorphs in a nuclear explosion. Nuclear warheads are stolen in *True Lies* (1994), Cameron's worst film. And the date of *The Terminator*'s Judgment Day references the Soviet Union's first detonation of an atomic bomb in 1949. As Cameron says, "The paranoia of [the *Terminator*] films is something very near and dear to my heart," explaining he's feared nuclear annihilation ever since seeing footage of nuclear explosions as a child. In its critique of military technology, Cameron boldly claimed *Terminator 2* would be "the first action movie advocating world peace."

In what might seem like unlucky timing, the film was released not too long after the Soviet Union collapsed and 'we' largely stopped worrying about imminent nuclear war. But it doesn't feel out-of-date, because it's really about how technological hubris (especially by men) leads towards doom, which is really the theme running through all of Cameron's work, so much so that the name of the club in which the Terminator locates Sarah in the first film, Tech-Noir, has become a shorthand for a dystopian technological future.

And so, John, Sarah, and the Terminator arrive at a seemingly abandoned ranch in the desert, actually the hideout of her friend Enrique, and where Sarah has stashed her gear, including lots of weapons and ammunition. They plan to cross the Mexican border.

Sarah falls asleep and has a nightmare in which she's trying to warn people about the apocalypse, but no-one hears her. A nuclear bomb explodes, incinerating Los Angeles. She wakes up in shock and sees the words she's carved into a table: 'NO FATE.' She grabs a rifle and tears off in a car. John recognizes it's the message from his future self, given to Sarah by Kyle: "The future is not set. There is no fate but what we make for ourselves." The Terminator realizes Sarah is going to kill Dyson. They take off after her.

At Dyson's house, Sarah snipes at Dyson from a distance, wounding but not killing him. She bursts into the house, and only stops herself from finishing Dyson when his young son pleads for his life. Sarah breaks down in tears. John and the Terminator arrive. The Terminator attends to Dyson's wound, then peels back its synthetic skin to reveal its metal arm—identical to the one Cyberdyne Systems keeps in a special vault at its lab.

"Dyson listened while the Terminator laid it all down. Skynet, Judgment Day, the history of things to come. It's not every day that you find out you are responsible for 3 billion deaths. He took it pretty well." Sarah is being sarcastic: Dyson nearly throws up.

In a deleted scene from the end of the first film, we learnt that parts of the Terminator were salvaged from the factory where Sarah and Kyle finally defeated the machine. The factory was owned by Cyberdyne Systems. As Dyson explains, this 'inspired' his revolutionary chip design. Within three years, Cyberdyne Systems became the largest supplier of military computer systems. And so, technology was accelerated through a time loop (at the risk of tempting fate, this is a nice counter to singularity believers: the only way 'we' created advanced A.I. is because it traveled back through time to reach us).

Dyson defends himself: he couldn't have known what it would lead to. Sarah explodes: "Fucking men like you built the hydrogen bomb. Men like you thought it up. You think you're so creative. You don't know what it's like to really create something, to create a life, feel it growing inside you. All you know how to create is death and destruction."

Now he knows the future, Dyson promises he'll stop working on the processor and quit his job, but Sarah and the Terminator convince him that nobody must be allowed to finish his work. The Terminator arm and chip held at Dyson's lab must be destroyed.

"IT HAS TO END HERE"

Miles Dyson is another example of 'Haven't you ever seen a sci-fi film?' – of not considering how new technologies might be used, of assuming they're only benign, and anyway, 'progress' is unstoppable, right? Earlier, when Dyson's colleague asks where the arm came from, Dyson gives him the same answer he got from his superiors: "Don't ask."

Worse than that, we invent technologies we know will be used for harm. So of course the U.S. National Security Agency named an A.I. drone program after Skynet, and the Chinese government has done the same for a video mass surveillance system, Tiānwǎng (天网). Too often, inventors lack a sufficient sense of future responsibility, failing to interrogate the possible impact of their technologies. The Californian Ideologists of big tech are only the most recent examples (naturally, Cyberdyne Systems is based in Silicon Valley). They speed us blindly towards the iceberg (as Cameron depicted in *Titanic*), or as Benjamin Noys puts it, help to ensure our "acceleration into the abyss." But unfortunately, as Mike Ryan notes, in Sarah and John's world, at least the 'inventor' of future destruction comes to recognize his culpability, whereas in our world, the Miles Dysons don't seem to care.[3]

As they head to Cyberdyne Systems, Sarah reflects, "The future, always so clear to me, had become like a black highway at night. We were in uncharted territory now, making up history as we went along." They arrive at the lab. Dyson tries to convince the security guard he just wants to show his 'friends' around, but the guard asks for authorization, so they lock him in the restroom. Another guard finds his colleague and raises the alarm.

John, Sarah, the Terminator, and Dyson break into the lab and start destroying computers and files, while the police arrive in force outside. John and Dyson open the vault and grab the metal arm and processor chip. Now they've got "Skynet by the balls." They rig the lab with explosives, while the Terminator fires a minigun and grenades to push back the police. A heavily armed SWAT team bursts in and opens fire. Dyson is severely wounded, and sacrifices himself by blowing up the lab while the others escape. However, the T-1000 also arrives on the scene (he's been monitoring police channels), 'procures' a police helicopter, and gives chase to John, Sarah, and the Terminator's stolen SWAT van.

The *Terminator* films play with the idea that the past-future can be changed. On first viewing, *Terminator 2*'s action is so spectacular and the special effects so unprecedented we forget to ask the obvious question: if by the end of the film Sarah and John will have succeeded in destroying the knowledge that leads to the machines, why don't the T-1000 and the Terminator instantly disappear? Indeed, why is there a story at all?

The future seems inevitable, since the killer robots have come back from it already. And yet there's John's message from the future. So is there any way to avoid the "accelerationist dystopia of The Terminator" (as Robin Mackay and Armen Avanessian put it in *#Accelerate: The Accelerationist Reader*)? Can there be 'no fate but what we make'?

There are tons of interesting, and mind-bending, explanations about why time travel stories like *The Terminator* may or may not make sense. But we mentioned the Novikov self-consistency principle in chapter 3 – that 'whatever happened, happened.' In *Terminator 3: Rise of the Machines*, we learn that Judgment Day was only delayed: a company called Cyber Research Systems purchased patents from Cyberdyne Systems, which had back-up systems, and the U.S. Air Force contracted it to develop a new cyber-defense system. The timeline continues to change in later stories from the series, creating a multiverse of temporal phenomena. But one way or another Skynet seems inevitable.

The argument here is that the past, and so the future (at least in its fundamentals), can't be changed. The new right aimed at stopping an alternative future, and they (largely) succeeded. Whatever we do now, we can't avoid 'judgment day.' In some sense, it's already happened. Instead, we need to prepare and organize the human resistance. As John Connor says in *Terminator 3*, humanity's "destiny was never to stop Judgment Day; it was merely to survive it."

"THIRTY-FIVE YEARS FROM NOW, YOU REPROGRAMMED ME TO BE YOUR PROTECTOR HERE, IN THIS TIME"

An incredible freeway chase follows. In the van, Sarah protects John under a pile of bulletproof vests, then fires on the T-1000's rapidly closing helicopter. The van passes under a freeway bridge, but the helicopter follows (they actually did this stunt in real life). Sarah is shot in the leg. The Terminator stops the van, causing the helicopter to collide into the back of it. The helicopter crumples and crashes onto the freeway, but the T-1000 emerges from the wreckage and swaps to a large truck with 'LIQUID NITROGEN' written on its side.

The apparent paradox of Cameron's films is that he uses, and has actually invented, groundbreaking technologies, but as we've discussed, he's deeply concerned about the unthinking use of technology, and questions the ideas underpinning industrial civilization. "Within us," he says, "we have both a compassionate sensitivity and a violent beast. That beast, coupled with technology, got us to where we are today and enabled us to dominate the planet" (unlike the accelerationists, he doesn't mean that in a good way).

Terminator 2 represents a major inflection point in filmmaking technology, reflected in the conflict between the practical effects original model Terminator and the pioneering computer-generated advanced T-1000. It's a near-perfect blend of both, before CGI took over and started making everything feel, well, artificial—which was arguably a problem with Cameron's *Avatar* (2009). Perhaps he should have followed his fictional human resistance. John Connor's Tech-Com army use and adapt technology, but their mission is to avoid being subject, literally slaves, to it. They use the machines' technology against them, and for human survival.

In his presentation *Terminator vs. Avatar*, Mark Fisher used Cameron's films to reflect on accelerationism. Noting the irony of employing cutting-edge technology to enable us to "play at being inner primitives" (in the case of *Avatar*), Fisher suggested that the only direction is 'forward.'[4] But these aren't the only options. *Terminator 2* says we might be able to avoid the machine dystopia not by abstaining from technology, but by avoiding *techno-utopianism*. Fisher seems to fall into the trap that's common among accelerationists (and the acceleration curious): caricaturing critics as hapless, even hypocritical, hippies. For example, Srnicek and Williams dismiss what they call the "folk politics" of the contemporary left, stemming from 1960s new left environmentalism and new social movements—what they see as the predilection for 'horizontalist' organization, direct democracy, spontaneous direct action, and the privileging of local concerns over global ones.

The problem, to Srnicek and Williams, is that these tactics don't 'scale.' They may work well enough in particular instances, but they've produced relatively little in the way of lasting change, including social structures and institutions. So they're not much help for building a larger and longer-enduring resistance, one that could significantly improve people's lives, let alone change (or save) the world.

But we discussed our concerns about a left accelerationist future in the previous chapter. The assumption that 'scale' is the only answer risks leading us further into the dark future of passivity, resignation, and anger among ordinary people, and bureaucracy, irresponsibility, and unaccountability among a technological elite class. Solutions to the crises we face can't (only) come from above. The idea that they can *is* the crisis.

"ARE YOU EVER AFRAID?"

Maybe left accelerationists should watch more time-travel films. As Ville Kellokumpu notes, "'fully automated luxury communism' seems to see history only in linear terms (we can only progress or regress), whereas crises such as climate change require us simultaneously to 'go backwards' by decelerating our consumption of resources, sideways by reconfiguring our relationship to nature, and forwards by investing in what really improves human existence."[5]

Srnicek and Williams, and others, might be right to challenge us about how local action can grow. But in survival, and given the improbability and dangers of left accelerationism, their challenge looks rather different: not as a reason to reject localism, more as a need to 'accelerate' our investment in it. In collapse, our current systems of industrial agriculture and global exchange, among other things, will suffer from more breakdowns. The local, horizontal, human scale, and communal will be critical to survival, not distractions from it.

'Accepting' collapse means building our own, much more secure social infrastructure. Radical decentralization and re-localization represent the only viable non-authoritarian alternative to disaster capitalism and emergency centralism, and the best, really the only, foundation for a new left politics of collapse.

This just recognizes what elites have known for some time. From the perspective of the rich and powerful, society has already collapsed. It collapsed ideologically in the 1970s and 1980s, when the new right promoted the idea that elites have no responsibility to anyone else, let alone the future. Practically it's been collapsing ever since. Now, they're literally building their bunkers. We need to build ours. Another end of the world is possible, to quote the title of Pablo Servigne, Raphaël Stevens, and Gauthier Chapelle's book.[6]

As part of this, survival-oriented local movements, organizations, and communities need to reclaim technology from big corporations and the state. One example is in community-owned local internet mesh networks, which will also be critical in collapse, for instance they can provide text messaging for communities when there are power outages. Recall the hippie history of early internet subculture, before its corruption by corporations. Rather than a naïve view that technology will naturally open-up possibilities for a post-capitalist future, we need to remember that technology has often been developed and used for capitalist purposes against workers. As sci-fi writer Ted Chiang says, most fears about technology are really fears about how capitalism will use technology—hence the long history of resistance to it by workers, and why we need a twenty-first-century worker resistance as well, as Gavin Mueller explains in *Breaking Things at Work* (2021).

At the very least, we need more democratic control over how technology is developed, as Lizzie O'Shea argues in *Future Histories* (2019). But the best way to ensure technology is used in socially positive ways is to own it. Among others, Trebor Scholz and Nathan Schneider in *Ours to Hack and to Own* (2017) call for a new kind of online economy, 'platform cooperativism,' combining the rich heritage of cooperatives with the promise of twenty-first-century technologies, free from monopoly, exploitation, and surveillance.

"OH, THIS IS DEEP"

And not just in communication technologies, but also energy, waste, agriculture and food, education and training, welfare, health services, mental health services, social care, childcare, mutual credit systems, alternatives to policing, alternative media . . . communities progressively 'seceding' from the dying machine's structures, systems, and controls.

As Helena Norberg-Hodge states, globalization in its current form has no future, but the good news is that, under the radar of mainstream media, the local future is already happening—a worldwide localization movement is emerging through which people are coming together to rebuild connections to one another, all the while repairing fractured communities and damaged environments.[7] The future isn't in as-yet-invented technologies coming to save us, it's in new (and old-as-new) ways we can organize ourselves. As Bill McKibben says, the technology we need most is, "The technology of community—the knowledge about how to cooperate to get things done." Or as Kim Stanley Robinson puts it, "social justice as survival technology."

Talking of survival, we're fast approaching the climax of the film. The T-100's truck bears down on the stolen SWAT van and tries to crush it. Now at the wheel, John steers the van into a steel mill. The Terminator climbs onto the hood of the T-1000's truck and turns the wheel, causing the truck to flip over and screech to a halt. Liquid nitrogen pours from the truck. The nitrogen freezes and then shatters the T-1000 – but only temporarily. The liquid metal monster re-forms and continues to pursue them.

John, Sarah, and the Terminator head deeper into the maze of the mill. They're trapped on the edge of a molten metal pool. The T-1000 approaches. The Terminator tells John and Sarah to flee, then fights the T-1000, which traps the Terminator's arm in a giant gear. The Terminator tears off its own arm and reengages in the fight. But the T-1000 crushes the Terminator with an iron bar, then spears it with a metal rod. The Terminator 'dies.' Now the T-1000 is finally free to finish off John and Sarah.

It's the end of hope, or what desperate hope John and Sarah were clinging to. One way to understand the (often derided) radical movements in Western democracies of the past few decades—from the anti-corporate/anti-globalization movement and Occupy, up to today's climate justice campaigns—is as potential worlds coming back from the future. And from the past as well, since these movements' thinking about possibilities, and their inclination to humane, holistic ways of living, working, and organizing echo their (as we've seen) predecessors in new left and environmental movements half a century earlier.

But as we've also discussed, recognizing collapse means there's less point protesting against a system that's disintegrating and revealing the vicious exterminator beneath its synthetic skin. Demonstrating defiance is important, but it also defers to authority to 'make better choices,' choices we would be better making for ourselves. As Nafeez Ahmed puts it: "No amount of nonviolent resistance will provide our existing political institutions with the capacity to address the crisis. Because the problem runs much, much deeper. . . . This is what is missing from our response to the crisis of civilization."[8]

Building local organizations, institutions, and networks for survival helps to bring another future (one possible future, anyway) into the present. The answer to crisis lies not in lists of complex policy proposals or theoretical language, but in practical measures of security. These are not new ideas, just largely unrealized ones. And again, it's not about making a choice between an accelerated future and a preindustrial past, but as James Robertson wrote in 1978 about a 'Sane, Humane, Ecological (SHE) Future', changing direction away from pursuing economic expansion beyond feasible limits, and so actually having a *better* chance of meeting economic needs and achieving a high quality of life for all.[9]

No, it's not as immediately attractive as fully automated luxury communism and its promised cornucopia of commodities. But we'll soon (be forced to) realize, as David Fleming put it, degrowth is not really a choice anyway—it's coming as a result of the climate crisis, and the only issue is whether we are ready for it or not.[10] Or as *The Limits to Growth*'s Jay Forrester says of his still-vocal critics, "I don't really expect to convert them. The only option is to outlive them."

"AREN'T WE CHANGING THINGS, I MEAN RIGHT NOW, CHANGING THE WAY IT GOES?"

After torturing Sarah to try to draw John out of hiding, the T-1000 morphs into her form. He tries to trick John, but Sarah shoots the T-1000. She runs out of shotgun rounds. The Terminator reappears—it's re-booted itself—and uses his grenades to blast the T-1000 into the molten pool. The T-1000 struggles and splashes until it's completely dissolved.

The Terminator confirms the T-1000 is finished. John throws the metal arm and processor chip into the molten pool. Sarah sighs that it's over. But the Terminator reminds her there's one more chip which needs to be destroyed: it points to its own head. John protests tearfully, but the Terminator insists they must melt it down as well. They lower the Terminator into the molten pool. The last thing we see is a final thumbs-up as the Terminator sinks completely into the bubbling metal lake. Its internal processing screen blacks out.

In terminating the Terminator, John and Sarah resist the accelerationist tendency. However tempting it might be, they don't fool themselves that *they* will only use the technology for good. Like them, we need to bring the future into the present by making decisions today as if they'll determine whether we'll live in dystopia or not (or maybe more realistically, how bad it will be), because in sum, they do. Unlike Dyson (before his awakening at least), we need to consider what world each of our decisions, including what we consume and what we produce, helps to bring into being. We can't avoid collapse, but we can help to decide whether it's truly catastrophic and dystopian, or maybe something else. And for this, we can imagine we're in the future already, then ask what we could have done to resist dystopia earlier and to build structures that help us to survive through collapse. It's a way, however limited it might seem individually, of regaining our power to create the future.

As time travel stories teach us, we can never really know if or how our actions in the present will bring about a better future. But we have to try, in the recognition the final battle will not be fought in the future, it will be fought here, in our present . . . *tonight.*

This was a core new left idea of course—of authenticity, responsibility, and defending humanity against the machine. The environmental new left was the first contemporary utopian-dystopian political movement, in the sense that it saw the future was at stake, and that personal everyday actions as well as macro-political decisions mattered. It thought in terms of systems, their consequences, and the human (or inhuman) values that underpin them. And it identified a madness, a sickness, in the exterminism of our current system.

So, in every action we can choose to support and fund utopia, and challenge and "defund dystopia" (to put it in the apt phrase of Sytonia Reid of *Green America* magazine). This matters, even in collapse, especially in collapse. As we've discussed, economic and environmental crises will provide the rationale for authoritarian responses from states and corporations, particularly against marginalized communities, to protect wealth and property. As the environmental activist Tim DeChristopher argues, "when things get ugly, and access to resources becomes difficult, we want to have trust that those making decisions will act justly, and not just favor the strong. . . . We need to start working now on putting in place power structures that share our values as we enter difficult times."[11]

"THE DELUSIONAL ARCHITECTURE IS FAIRLY UNIQUE"

Fredric Jameson makes the point in *Archaeologies of the Future* (2007) that while the opposite of utopia is dystopia, the force that opposes utopia is better understood as 'anti-utopia.' As we've discussed, dystopian fiction isn't necessarily against the idea of utopia, more often it identifies trends in the present that, unchecked, will lead to disaster. But anti-utopianism says even the idea of utopia, let alone programs for it, are dangerous.

We've also been skeptical of utopias here, particularly supposedly universal schemes for perfection. But as discussed in previous chapters, it's the totalizing nature of some ideologies that makes them utopian-dystopian, not the idea of social improvement itself.

The new right was an anti-utopian project, and also (hypocritically, and in disguise) utopian-dystopian. It said: all left or even liberal hopes and policies for a better, more rational, fairer society, are not only doomed, but dangerous. But it also said: we've discovered the perfect machine that can create paradise, but only if we allow it to, only if we grant it near-absolute power.

If we're going to find hope, we need to reclaim the idea of utopia, but leave behind totalizing thoughts and desires. Jameson calls this 'anti-anti-utopianism.' It's the insistence it's possible to make a much better world but not a perfect one, since that's unrealizable: "[U]topias are non-fictional, even though they are also non-existent. Utopias in fact come to us as barely audible messages from a future that may never come into being. I leave the articulation of that message to Marge Piercy's Mattapoisett utopians—time-travellers from a future which, they warn us, without ourselves and our own present, may never come into existence."[12]

'Mattapoisett utopians' is a reference to Marge Piercy's 1976 novel *Woman on the Edge of Time*, now considered a classic of utopian speculative science fiction as well as a feminist classic. *Terminator 2* even echoes it in some respects. The story concerns a poor Mexican-American woman named Consuelo (Connie), who is unjustly incarcerated in a New York mental hospital due to her supposed violent criminal tendencies. But Connie has been receiving communications from a time-travelling figure from the future, Luciente, who tells her about the year 2137, in which pollution, patriarchy, homelessness, homophobia, racism, ethnocentrism, sexism, class-subordination, imperialism, and totalitarianism have been eradicated (but only after the existing civilization had been destroyed through environmental degradation and war). This society resembles a loose version of anarchism, in which many of the goals of the 1960s and 1970s radical movements have been fulfilled.

However, Luciente also shows Connie alternative futures, including a consumerist, hyper-capitalist, environmentally sick, and extremely classist, racist, and gender-stratified society in which the wealthy elite live on space platforms, sustain themselves by dominating and exploiting the majority of the population through controlling knowledge and technology, harvesting human organs, and determining the masses' moods using psychotropic drugs.

Piercy's novel doesn't decide for the reader whether Connie's visions are the result of mental instability or not, but her confrontation with the future inspires her to revolt against her institutional captors. She comes to feel she's living at an important time in history, and that her actions and decisions will determine the course of history. She heroically uses her limited means to try to prevent the dissemination of the mind-control technology that will make the future dystopia possible, putting an end to the mind-control experiments and lobotomies planned for hundreds of her fellow imprisoned patients. She fights back.

"IN AN INSANE WORLD, IT WAS THE SANEST CHOICE"

Whatever our pessimism, there are some positive futures already in the present. If we look, we can see we're already surrounded by what Erik Olin Wright calls 'real utopias,' enclaves of anti-capitalism within capitalism, from worker cooperatives to open-source projects and other experiments in economic democracy. And, as Conor Owens further suggests, while most people might not be able to picture a far-off future beyond scarcity and hierarchy, they can envisage a more immediate future which contains more of the kinds of (rather anarchistic, in the political sense of the word) things they can see for themselves every day, such as participatory budgeting programs, popular neighbourhood assemblies, and much else.[13]

Of course, this is the kind of prefigurative argument that left statists and accelerationists typically disparage. But the Marxist philosopher Ernst Bloch saw utopian thinking as governing everything future-oriented in life and culture. This is not to see utopia as a program that can be implemented only through a closed system that can easily slip into absolutism. Instead, Bloch urged us to see utopia as an impulse, and as a continual process of approximating an ideal in day-to-day life.

Bloch's great book was called *The Principle of Hope* (published in three volumes from 1954 to 1959), which examined the ways that popular culture, literature, philosophy, religion, and even daydreams, often dismissed as mere ideology by other Marxists, can contain emancipatory moments which project visions of a better life and challenge capitalism. Bloch believed that left thinking and politics should heed these visions to provide programs and arguments which appeal to the deep desire for a better life held by everyone. Or as Kim Stanley Robinson says, we shouldn't define 'utopia' as a perfect end-state society, rather it's "a process, not a product . . . just a name for one kind of history."[14]

Of course, this history-in-process was exactly what the new right came to terminate. But it didn't succeed, not completely. It promoted cynicism and nihilism, but as Sandra Newman counters: "The world is dying, yet we're still being scolded that we must not try to do better. . . . We must leave behind the superstition that every attempt to solve our problems ends in Orwellian dystopia. . . . If we are to do better, we must free ourselves to try."[15]

"MY MISSION IS TO PROTECT YOU"

In this book, we've draw on dystopian depictions of the future. But popular culture largely lacks realistic but hopeful visions of the better worlds we could, and have to, build. And as Alex Steffen says, "You can't build what you can't imagine."

Fredric Jameson suggests that there have been no great utopian texts since the widespread introduction of computers, the last being Ernest Callenbach's *Ecotopia* published in 1975, which describes an ecological utopia and was influential on the counterculture and green movement in the 1970s and after. Instead, we've just had the "free market deliria" of cyberpunk, which assumes that capitalism is itself a kind of utopia of difference and variety.[16]

And so, even though their visions of the coming future remain accurate in many ways, we need to move on from genres such as cyberpunk. There's no point warning us anymore about a future 'we' chose not to try to avert. Mark Fisher said we need new "economic science fictions." But we won't find them in accelerationist fantasies (or nightmares). Instead, they're in climate-oriented science fiction ('cli-fi') and in developing genres such as solarpunk. These inspire a postcollapse, but postnihilist, imagination.

As Sarena Ulibarri describes it, solarpunk is a method of imagining a better future, based around sustainable technologies and cooperative, rather than competitive, living. Solarpunk is a literary and visual movement that originated in Brazil in the early 2000s, captured in the science fiction short story collection, *Solarpunk: Histórias Ecológicas e Fantásticas em um Mundo Sustenavel* (Ecological Fantastic Stories in a Sustainable World), published in Brazil in 2012.

It's 'punk' because solarpunks rebel by refusing to accept that disaster and dystopia are the only possible futures. According to Adam Flynn's *Solarpunk Manifesto*, the genre is inspired by "ingenuity, generativity, independence and community." In place of cyberpunk certainty, solarpunk imagines the future we might have if we took the alternative route: replacing states and corporations which free federations of communities and cooperatives, and using decentralist and ecological technologies to create a world beyond economic scarcity and social hierarchy, defined by autonomy, mutual aid, diversity, and inclusiveness. Flynn suggests the movement begins with "infrastructure as a form of resistance."

Echoing the human resistance (not a bad name for a future anti-dystopian movement), solarpunks draw on existing technologies, like solar energy and urban agriculture, to "remix the present to produce an alternative future", as Jennifer Hamilton puts it.[17] Again, the real challenge is not in technological re-tooling, but in the societal reworking necessary to make a carbon-free, fair future viable.

Solarpunk is aspirational, yes, but it's more realistic than the accelerationist alternative. It's a practical utopianism, rather than the as-yet-invented miracle techno-cures of neoliberalism, or of centrist or left accelerationism. As Hamilton says, it's radical, but not radically impossible. Many or most of the technologies and practices that solarpunks draw on already exist: solar and other renewable energy, urban agriculture, or organic architecture and design. . . . So do the social practices.

If solarpunk has a 1970s vibe to it, that reflects a resurgence of the recognition of the limits to growth, the need for new ways of living and organizing, and the importance of hope in the context of coming collapse. It also resuscitates the idea of "utopian centers of new cultural experiment," in Tom Hayden's phrase (Hayden was a major new left figure). For example, in the 1970s he proposed 'free territories,' urban neighborhoods or rural communes where people could collectively fashion a new social order in contrast to mainstream society. These would act, in libertarian socialist terminology, as centers of 'dual power' or 'counterpower,' the nonviolent strategy of achieving a libertarian socialist economy and polity by incrementally establishing and then networking institutions of direct participatory democracy to contest the existing power structures of state and capital.

This doesn't necessarily mean disengagement with existing institutions, but it does mean building our own alternatives. Solarpunk then belongs to the longer tradition of the decentralist left, of thinkers and activists such as Peter Kropotkin, William Morris, Emma Goldman, Lewis Mumford, Paul Goodman, E. F. Schumacher, and Murray Bookchin. But it also thinks forward, to how new technologies might be used for collective survival.

"WE STILL HAVE TO STOP THIS FROM HAPPENING, DON'T WE?"

Sarah holds the sobbing John to her shoulder. She's finally able to (re-)connect with her son, and with her own compassion, caring and nurturing—her own humanity.

We're told at the beginning of the film that Judgment Day occurs on 29 August 1997. Then, because of John, Sarah, and the Terminator, it's delayed—but only delayed. Sarah's opening monologue is about the year 2029, so whenever Judgment Day actually happens, she's talking about a future war that's been raging for some time.

We can find ways to survive in the dystopias to come, maybe for a long time—if we prepare for them now. And remember, the machines sent the

Terminators back in time to our present because the human resistance, under John's leadership, was finally *winning*.

Subsequent films in the *Terminator* series have been decreasingly popular. The latest, and possibly the last film, *Terminator: Dark Fate* (2019), directed by Tim Miller, was a box office failure, and received very mixed reviews (somewhat unfairly, I'd suggest). After a run of weak entries, failed attempts to reframe the series, and an increasingly convoluted timeline, maybe the Terminator really is, finally, dead. Generationally as well, it's difficult to get younger people engaged in a story, born of Cold War nuclear nightmares, that began nearly forty years ago. We need new stories for a new future.

At the beginning of *Dark Fate*, John is finally killed by the Terminator. Many years later, in a nice twist, Sarah is hunting Terminators, which are still being sent back from the future. What she doesn't know is *why*. It turns out there are other leaders, and other futures. Some people will always find ways to resist the machines, and change the future. As John Connor said, "All I know is what the Terminator taught me: Never stop fighting."

NOTES

1. Derrick Jensen, Aric McBay, and Lierre Keith, *Deep Green Resistance: Strategy to Save the Planet* (New York: Seven Stories Press, 2011).

2. Natalie Osborne, "For Still Possible Cities: A Politics of Failure for the Politically Depressed," *Australian Geographer* 50, issue 2 (2019).

3. Mike Ryan, "'Terminator 2: Judgement Day' Is Relevant Again Today and That Sucks," *Uproxx*, January 3, 2018. https://uproxx.com/movies/terminator-2-judgement-day/.

4. Mark Fisher, "Terminator vs. Avatar: Notes on Accelerationism," Mark Fisher blog, presented at the Accelerationism symposium, Goldsmiths, University of London, September 14, 2010. https://markfisherreblog.tumblr.com/post/32522465887/terminator-vs-avatar-notes-on-accelerationism. Also Mark Fisher, "Terminator vs. Avatar," in *#Accelerate: The Accelerationist Reader*, Robin Mackay and Armen Avanessian (eds.) (Falmouth: Urbanomic Media Ltd, 2014), 335–46.

5. Ville Kellokumpu, "Fully Automated Luxury Communism: A Manifesto By Aaron Bastani," *Society & Space*, September 3, 2019. www.societyandspace.org/articles/fully-automated-luxury-communism-a-manifesto-by-aaron-bastani.

6. Pablo Servigne, Raphaël Stevens, and Gauthier Chapelle, *Another End of the World is Possible: Living the Collapse (and Not Merely Surviving It)* (Cambridge: Polity, 2021).

7. Helena Norberg-Hodge, *Local Is Our Future: Steps to an Economics of Happiness* (Totnes: Local Futures, 2019).

8. Nafeez Ahmed, "Paradigm Shift Is Needed to Address Crises of Civilization," *Mondoweiss*, May 14, 2019. https://mondoweiss.net/2019/05/paradigm-address-civilization/.

9. James Robertson, *The Sane Alternative, A Choice of Futures* (St. Paul, MN: River Basin Publishing Co., 1978/1983), 13, 36. www.jamesrobertson.com/book/thesanealternative.pdf.

10. David Fleming, quoted in the "Foreword" by Jonathon Porritt, in *Lean Logic, A Dictionary for The Future and How to Survive It*, David Fleming (White River Junction, VT: Chelsea Green Publishing, 2016).

11. Quoted in Nick Buxton, "Time to be Honest," *Red Pepper*, May 25, 2011. https://www.redpepper.org.uk/time-to-be-honest/.

12. Fredric Jameson, "The Politics of Utopia," *New Left Review* 25 (January–February 2004), 35–56.

13. Conor Owens, "Social Anarchist Futures," *Solarpunk Anarchist*, May 2, 2018. https://solarpunkanarchists.com/2018/05/02/social-anarchist-futures/.

14. Tasha Robinson, "We Asked Kim Stanley Robinson: Can Science Fiction Save Us?," *Polygon*, October 20, 2020. www.polygon.com/2020/10/20/21525509/kim-stanley-robinson-interview-science-fiction-utopias.

15. Sandra Newman, "The Other Side of Black Mirror: Literary Utopias Offer the Seeds of Better Real Life," *The Guardian*, May 14, 2019.

16. Fredric Jameson, "In Soviet Arcadia," *New Left Review* 75 (May–June 2012): 119–27.

17. Jennifer Hamilton, "Explainer: 'Solarpunk', or How to be an Optimistic Radical," *The Conversation*, July 20, 2017. https://theconversation.com/explainer-solarpunk-or-how-to-be-an-optimistic-radical-80275.

Conclusion

"Mom! We Need to Be a Little More Constructive Here, Okay?"

At the end of *Terminator 2*, we're travelling down a highway at night, again. Sarah says, "The unknown future rolls towards us. I face it for the first time with a sense of hope. Because if a machine, a Terminator, can learn the value of human life, maybe we can too."

As Jonathan Lear writes in *Radical Hope* (2006), "What makes . . . hope radical is that it is directed toward a future goodness that transcends [our] current ability to understand what [that] is." In every moment, there are always routes to different futures.

Terminator 2 originally ended even more optimistically. We're in a playground again, but thirty years in the future (in other words, just about now). Sarah watches children playing, including an adult John playing with his own daughter:

> August 29, 1997, came and went. Nothing much happened. . . . People went to work as they always do. Laughed, complained, watched TV, made love. I wanted to run to through the street yelling to grab them all and say, "Every day from this day on is a gift. Use it well." Instead, I got drunk. . . . But the dark future which never came still exists for me. And it always will, like the traces of a dream. John fights the war differently than it was foretold. Here, on the battlefield of the Senate, his weapons were common sense and hope. The luxury of hope was given me by the Terminator. Because if a machine can learn the value of human life, maybe we can too.

In contrast to some of the other films we've discussed, it was the studio that wanted a more ambiguous ending. Mario Kassar, the lead producer, felt it was too neat, too positive compared to the dark tone of the rest of the story, as if it came from a different film (a more open ending, implying that the struggle for the future is an ongoing one, would also more easily allow for sequels). Test audiences agreed, and James Cameron relented.

This book has been less about how dystopias have warned us about some distant, possible future, and more what they've told us about the fast-oncoming, increasingly probable present. Since we're here now, it might be time to move on from fictional dystopias. Obviously, good sci-fi continues to be produced, and consumed, as it should. But, we might ask, what do they really tell us that we don't already know—about ecological catastrophe, class exploitation, and technological hubris? Just as the films highlighted here predicted, the future is becoming less unknown. Paradoxically, 'no fate but that we make is right' is correct—but only if we first accept the reality of the world, and of the world soon to come.

This doesn't mean it's over. There might be no future emergency for which we must prepare. Dystopia is here already for some people—it has been for a long time—and for more of us soon. Magical technologies won't save us, nor will charismatic leaders touting simplistic solutions. But creativity, communities, and commitment will help to save some of us, in some times and places. Just how many times and places, and how many people, is still up to us.

And while much of the argument here has been about inevitability, about the unavoidability of crisis and collapse, there are also better things we can't anticipate, maybe can't even begin to imagine. There are the things we don't yet know—ideas, organizations, movements, ways of living and working and more—in the minds and hearts of new generations who are growing up in collapse and (yet) still want to build better, safer, fairer futures.

In *Terminator: Dark Fate*, Grace (Mackenzie Davis), a human augmented future super soldier, grabs Sarah's phone to figure out the geographical coordinates of an anonymous text message sent to the device. "What are you doing?", asks Sarah. Not wanting to waste time trying to explain her advanced knowledge, Grace responds, "Future shit."

Author Afterword

"I Need a Vacation"

Thank you for buying *Come With Me If You Want To Live*, and for making it to the end—as it were.

There really was a *Book of the Future*, and it really was one of my favorites when I was growing up. I distinctly recall the double page spread 'Two trips to the 21st century' and its stark contrast between a vibrant, green city, and a choking, dying world. I read it more than forty years ago, which seems crazy now. But I still remember drawing pictures of futuristic brutalist cityscapes based on the dying world. I guess that explains this book.

What also explains it are the books that precede it. After studying politics at university, I became a researcher, moving around different organizations every few years. I've been lucky enough to work in lots of different areas, from education and technology, to innovation, government, public services, and urban planning. The first book I wrote, *A Future for Planning: Taking Responsibility for Twenty-First Century Challenges* (Routledge, 2019), came out of urban planning. As the title suggests, that book was about the big challenges we face in this century, from climate change and demographic change, to automation and artificial intelligence, political instability and systemic risk. It made me think, if our politicians aren't preparing us for these problems, where is the book that tells young people how they might have to navigate a very different future from today?

But whenever I mentioned the idea to people, they'd express skepticism. Sure, they could see how it was needed, but wouldn't it be really *depressing* (which is to say, did I think anyone would buy it?). That made me think about the importance of hope, and the need for stories to promote a new sense of hope among young people . . . and out of that came my second book, *Welcome to The Rebellion: A New Hope in Radical Politics* (Zero Books, 2020). I never expected to write a book about politics and *Star Wars*, but the more I thought about it, the more it kind of kept working . . . and by then I'd pretty much written it.

I was part of the original *Star Wars* generation—*A New Hope* (1977) was the first film I saw at the cinema. The theme I used for my next book, *The Hunger Games*, didn't come out of my own generational experience. It's not that I didn't appreciate the films (I hadn't read the novels until I started working on the book). I *did* think they were strikingly serious, well-made, and acted. But *The Hunger Games* was never my story. Like *Star Wars*, though, it too was made for young people, today's millennials and then Generation Y, and clearly it spoke to them in some important ways. I came back to this when I was thinking about writing a new book: what stories today are the equivalent of what *Star Wars* was to me? And so that's how I ended up writing *Stay Alive: Surviving Capitalism's Coming Hunger Games* (Zero Books, 2021).

If *Star Wars* is about hope, in some respects *The Hunger Games* is about the lack of it. Most of its characters are fatalistic about the possibility of overthrowing their dystopian Panem (and yet they do—maybe). And, it seems, many young people today are pessimistic about how we'll face the crises that are likely to define their world in the twenty-first century. But a certain degree of hopelessness might be important; maybe it's more dangerous to be naive about the future that faces us now.

Which leads to this book, and reflecting on what we've been told for decades—really, for the whole of my life—about what's going to happen next, including, prominently, through films (which have always been my favorite form of entertainment). The connecting theme of course is the power of popular culture to say things that otherwise seem unsayable, whether about the need for revolutionary action against despotic power, or a fast-approaching fear-filled future.

Perhaps because in some way I realize now I've been thinking about all this for so long, the experience of writing the book hasn't been as depressing as you might assume. But I recognize that your experience might have been different. In which case, do make sure to take care of yourself. The point of this book isn't to induce fatalism, but rather a new realism—and from this, some new, maybe even exciting, ideas about what we do next.

If you have a few moments, please do leave a review (whether you liked the book or not) on the usual book-selling sites. I'd really appreciate it. And if you'd be interested in hearing about forthcoming books, do visit my website at: www.drmichaeljharrisbooks.com

And thanks again.

Bibliography

Ahmed, Nafeez. "Paradigm Shift Is Needed to Address Crises of Civilization." *Mondoweiss*, May 14, 2019. https://mondoweiss.net/2019/05/paradigm-address-civilization/

Anderson, Kurt. *Evil Geniuses: The Unmaking of America: A Recent History*. New York: Random House, 2020. Arthur, W. Brian. "The Second Economy." *McKinsey Quarterly*, October 1, 2011. www.mckinsey.com/capabilities/strategy-and-corporate-finance/our-insights/the-second-economy

Astle, Richard. "Rollerball Superstar, or Jesus Christ on Skates." *Jump Cut: A Review of Contemporary Media* 10–11 (1976): 23–24. www.ejumpcut.org/archive/onlinessays/JC10-11folder/Rollerball.html

Baker, Peter C. "'We Can't Go Back to Normal': How Will Coronavirus Change the World?" *The Guardian*, March 31, 2020.

Barbrook, Richard, and Andy Cameron. "The Californian Ideology." *Mute* 1, no. 3, 1 September (1995). www.metamute.org/editorial/articles/californian-ideology

Bastani, Aaron. *Fully Automated Luxury Communism*. London: Verso, 2019.

Berman, Marshall. *All That Is Solid Melts into Air: The Experience of Modernity*. New York: Verso, 1982.

Black, J. F. *The Greenhouse Effect*. Place unknown: Exxon Research and Engineering Company, June 1978. https://mk0insideclimats3pe4.kinstacdn.com/wp-content/uploads/2015/09/James-Black-1977-Presentation.pdf

Bould, Mark. "Dulltopia, On the Dystopian Impulses of Slow Cinema." *Boston Review* 22, January (2018). http://bostonreview.net/literature-culture-arts-society/mark-bould-dulltopia

Brin, David. *The Transparent Society*. New York: Basic Books, 1998.

Buxton, Nick. "Time to be Honest." *Red Pepper*, May 25, 2011. https://www.redpepper.org.uk/time-to-be-honest/

Carlson, Colin J., Gregory F. Albery, Cory Merow, Christopher H. Trisos, Casey M. Zipfel, Evan A. Eskew, Kevin J. Olival, Noam Ross, and Shweta Bansal. "Climate Change will Drive Novel Cross-Species Viral Transmission." *bioRXiv* (2020). www.biorxiv.org/content/10.1101/2020.01.24.918755v1

Catton Jr., William R. *Overshoot: The Ecological Basis of Revolutionary Change*. Urbana University of Illinois Press, 1980.

Cole, Jake. "Remake/Remodel: Rollerball (1975) vs. Rollerball (2002)." *Spectrum Culture* 28, April, 2013. https://spectrumculture.com/2013/04/28/remakeremodel-rollerball-1975-vs-rollerball-2002/

Claisse, Frédéric, and Pierre Delvenne, "Building on Anticipation: Dystopia as Empowerment." *Current Sociology* 63, no. 2 (2014): 155–69.

Csicsery-Ronay Jr., Istvan. *The Seven Beauties of Science Fiction*. Middletown, CT: Wesleyan University Press, 2008.

Curtis, Adam. "The Curse of TINA." *BBC*, September 13, 2011. www.bbc.co.uk/blogs/adamcurtis/entries/fdb484c8-99a1-32a3-83be-20108374b985

Curtis, Theo. "Whatever Happened to the Future? Lessons from the 1930s." *In The Long Run*, July 18, 2018. www.inthelongrun.org/articles/article/whatever-happened-to-the-future-lessons-from-the-1930s/

Davidow, William, and Michael Malone. *The Autonomous Revolution: Reclaiming the Future We've Sold to Machines*. Oakland, CA: Berrett-Koehler Publishers, 2020.

Davies, William (ed.). *Economic Science Fictions*. Cambridge, MA: MIT Press, 2019.

Davis, Mike. "The Coronavirus Crisis Is a Monster Fueled by Capitalism." *In These Times*, March 20, 2020. http://inthesetimes.com/article/22394/coronavirus-crisis-capitalism-covid-19-monster-mike-davis

Dean, Joan F. "Between 2001 and Star Wars." *Journal of Popular Film and Television* 7, no. 1 (1978): 32–41. www.tandfonline.com/doi/abs/10.1080/01956051.1978.9944190?journalCode=vjpf20

Ellis, Chris. *The Noah Virus: Who is Infected with High Resiliency for Disaster?* August, 2020. www.researchgate.net/publication/343363026_The_Noah_Virus_Who_is_Infected_with_High_Resiliency_for_Disaster

El-Sayed, Abdul. "Coronavirus is Exploiting an Underlying Condition: Our Epidemic of Insecurity." *The Guardian*, March 19, 2020.

Emerge. "Vinay Gupta – Waking Up in the Monster Factory (pt. 1)." *Emerge: Making Sense of What's Next* podcast, April 1, 2019. https://anchor.fm/emerge/episodes/Vinay-Gupta---Waking-Up-in-the-Monster-Factory-pt--1-e3ikvk

Eubanks, Virginia. *Automating Inequality: How High-Tech Tools Profile, Police and Punish the Poor*. New York: St Martin's Press, 2018.

Farrell, John. "Soylent Green: Cyberpunk's Horrifying Apotheosis." *Neon Dystopia*, December 1, 2020. www.neondystopia.com/cyberpunk-movies-anime/soylent-green-cyberpunks-horrifying-apotheosis/

Fisher, Mark. "Terminator vs. Avatar." In *#Accelerate: The Accelerationist Reader*, edited by Robin Mackay and Armen Avanessian, 335–46. Falmouth: Urbanomic Media Ltd, 2014.

Fisher, Mark. "Terminator vs. Avatar: Notes on Accelerationism." Mark Fisher blog, presented at the Accelerationism symposium, Goldsmiths, University of London, September 14, 2010. https://markfisherreblog.tumblr.com/post/32522465887/terminator-vs-avatar-notes-on-accelerationism

Fleming, David. *Lean Logic, A Dictionary for The Future and How to Survive It*. White River Junction, VT: Chelsea Green Publishing, 2016.

Frase, Peter. *Four Futures: Life After Capitalism*. London: Verso, 2016.

Frey, Carl Benedikt, Thor Berger, and Chinchih Chen. *Political Machinery: Did Robots Swing the 2016 US Presidential Election?*, 2018. www.oxfordmartin.ox.ac.uk/downloads/academic/Political%20Machinery_171008_CF5.pdf

Friedman, John, Michael S. Morris, Igor D. Novikov, Fernando Echeverria, Gunnar Klinkhammer, Kip S. Thorne, and Ulvi Yurtsever. "Cauchy Problem in Spacetimes with Closed Timelike Curves." *Physical Review D* 42, no. 6 (1990): 1915–1930. https://authors.library.caltech.edu/3737/

Fukuyama, Francis. "The End of History?" *The National Interest* 16 (1989): 3–18.

Gatland, Kenneth, and David Jefferis. *The Usborne Book of the Future*. London: Usborne, 1979.

Gibson, William. *Burning Chrome*. London: Harper Collins, 1986.

Glass, Fred. "Brazil (1985) By Terry Gilliam." *Film Quarterly* 39, no. 4, summer (1986): 22–28. https://scrapsfromtheloft.com/2019/03/12/brazil-terry-gilliam-review-fred-glass/

Graeber, David. "A Practical Utopian's Guide to the Coming Collapse." *The Baffler*, no. 22, April (2013). https://thebaffler.com/salvos/a-practical-utopians-guide-to-the-coming-collapse

Graeber, David. "Of Flying Cars and the Declining Rate of Profit." *The Baffler*, no. 19, March (2012). https://thebaffler.com/salvos/of-flying-cars-and-the-declining-rate-of-profit

Haiven, Max. "Capital's Vengeful Utopia: Unpayable Debts from Above and Below." *L'Internationale Online*, May 11, 2020. www.internationaleonline.org/research/real_democracy/131_capitals_vengeful_utopia_unpayable_debts_from_above_and_below/

Haiven, Max. *Revenge Capitalism: The Ghosts of Empire, the Demons of Capital, and the Settling of Unpayable Debts*. London: Pluto Books, 2020.

Hall, J. Storrs. *Where is My Flying Car?* San Francisco: Stripe Press, 2021.

Hamilton, Jennifer. "Explainer: 'Solarpunk', or How to Be an Optimistic Radical." *The Conversation*, July 20, 2017. https://theconversation.com/explainer-solarpunk-or-how-to-be-an-optimistic-radical-80275

Hari, Johann. *Lost Connections: Why You're Depressed and How to Find Hope*. New York: Bloomsbury, 2018.

Harris, Michael. *Stay Alive: Surviving Capitalism's Coming Hunger Games*. Winchester: Zero Books, 2021.

Hickman, Craig. "Fredric Jameson – Utopia, Dystopia, and the Myth of Neoliberalism." *Copyriot.com*, September 28, 2016. https://non.copyriot.com/fredric-jameson-utopia-dystopia-and-the-myth-of-neoliberalism/

Intergovernmental Science-Policy Platform on Biodiversity and Ecosystem Services (IPBES). *Global assessment Report on Biodiversity and Ecosystem Services of the Intergovernmental Science-Policy Platform on Biodiversity and Ecosystem Services*. Bonn: IPBES Secretariat, 2019.

Jacob, Jeffrey. *New Pioneers: The Back-to-the-Land Movement and the Search for a Sustainable Future*. University Park, PA: Pennsylvania State University Press, 1997.

Jacobs, Meg. *Panic at the Pump: The Energy Crisis and the Transformation of American Politics in the 1970s*. New York: Hill and Wang, 2017.

Jameson, Fredric. "In Soviet Arcadia." *New Left Review* 75, May-June (2012): 119–27.

Jameson, Fredric. "The Politics of Utopia." *New Left Review* 25, January-February (2004): 35–56.

Owens, Conor. "Social Anarchist Futures." *Solarpunk Anarchist*, May 2, 2018. https://solarpunkanarchists.com/2018/05/02/social-anarchist-futures/

Jensen, Derrick, Aric McBay, and Lierre Keith. *Deep Green Resistance: Strategy to Save the Planet*. New York: Seven Stories Press, 2011.

Kael, Pauline. "Baby, The Rain Must Fall." *The New Yorker*, July 12, 1982.

Kaplan, Robert D. "The Neo-Malthusian World of the Coronavirus, How the Natural Environment is a Force for Instability." *The National Interest*, February 28 (2020). https://nationalinterest.org/blog/buzz/neo-malthusian-world-coronavirus-127947

Kellokumpu, Ville. "Fully Automated Luxury Communism: A Manifesto By Aaron Bastani." *Society & Space*, September 3 (2019). www.societyandspace.org/articles/fully-automated-luxury-communism-a-manifesto-by-aaron-bastani

Kendzior, Sarah. *Hiding in Plain Sight: The Invention of Donald Trump and the Erosion of America*. New York: Flatiron Books, 2020.

Kessel, Jonah M. "How to Stop the Next Pandemic." *The New York Times*, September 1, 2020.

Kord, Susanne. "Pandemic Movies: What Terry Gilliam's '12 Monkeys' Tells Us About the Future." *Liverpool University Press Blog*, June 17, 2020. https://liverpooluniversitypress.blog/2020/06/17/pandemic-movies-what-terry-gilliams-12-monkeys-tells-us-about-the-future/

Kord, Susanne. *12 Monkeys*. Oxford/Liverpool: Auteur Publishing in partnership with Liverpool University Press, 2019.

Kriss, Sam. "Abandon the Future." *Jacobin*, November 21, 2015. www.jacobinmag.com/2015/11/back-to-the-future-day-marty-mcfly-sam-kriss

Lepore, Jill. "A Golden Age for Dystopian Fiction." *The New Yorker*, June 5 and 12, 2017. www.newyorker.com/magazine/2017/06/05/a-golden-age-for-dystopian-fiction

Lilley, Sasha, David McNally, Eddie Yuen, and James Davis. *Catastrophism: The Apocalyptic Politics of Collapse and Rebirth*. Oakland, CA: PM Press/Spectre, 2012.

Lloyd's, *Food System Shock, The Insurance Impacts of Acute Disruption to Global Food Supply*. London: Lloyd's, 2015.

Lyttelton, Oliver. "20 Oddball Sci-Fi Films of the 1970s." *IndieWire*, March 7, 2013. www.indiewire.com/feature/20-oddball-sci-fi-films-of-the-1970s-101018/

Mackie, Brendan. "Meritocracy and Battle Royale." *Los Angeles Review of Books*, December 10, 2020. https://lareviewofbooks.org/article/meritocracy-and-battle-royale/

Marohn, Charles. "The Growth Ponzi Scheme, Part 4." *Strong Towns*, June 14, 2011. www.strongtowns.org/journal/2011/6/16/the-growth-ponzi-scheme-part-4.html

Meadows, Donella, Jorgen Randers, and Dennis Meadows. "A Synopsis: Limits to Growth: The 30-Year Update." *donnellameadows.org*, undated. http://donellameadows.org/archives/a-synopsis-limits-to-growth-the-30-year-update/

Meadows, Donella, Jorgen Randers, and Dennis Meadows. *Limits to Growth: The 30-Year Update*. White River Junction, VT: Chelsea Green, 2004.

Meadows, Donella H., Dennis L. Meadows, Jørgen Randers, William W. Behrens III. *The Limits to Growth*. Washington, D.C.: Potomac Associates, 1972.

Mitchell Jr., Richard J. *Dancing at Armageddon: Survivalism and Chaos in Modern Times*. Chicago: University of Chicago Press, 2002.

Morello, Lauren. "Phytoplankton Population Drops 40 Percent Since 1950." *Scientific American*, July 29, 2010. www.scientificamerican.com/article/phytoplankton-population/

Mostafa, Joshua. "The Revolution Will Not Be Automated." *Sydney Review of Books*, July 23, 2019. https://sydneyreviewofbooks.com/review/zuboff-bastani/

Murphy, Douglas. *Last Futures: Nature, Technology, and the End of Architecture*. London: Verso, 2016.

National Research Council. *Carbon Dioxide and Climate: A Scientific Assessment*. Washington, D.C.: The National Academies Press, 1979.

Napper, T. R. "A Post-Pandemic World: The Ultimate Cyberpunk Primer Discussions." *Grimdark Magazine*, May 20, 2020. www.grimdarkmagazine.com/a-post-pandemic-world-the-ultimate-cyberpunk-primer/

Napper, T. R. "Kim Stanley Robinson Is Wrong About Something." *T.R. Napper*, April 5, 2019. www.nappertime.com/kim-stanley-robinson-wrong-something/

Newman, Sandra. "The Other Side of Black Mirror: Literary Utopias Offer the Seeds of Better Real Life." *The Guardian*, May 14, 2019.

Norberg-Hodge, Helena. *Local Is Our Future: Steps to an Economics of Happiness*. Totnes: Local Futures, 2019.

Nordhaus, Ted. "The Empty Radicalism of the Climate Apocalypse." *Issues in Science and Technology* XXXV, no. 4, summer (2019): 69–78. https://issues.org/the-empty-radicalism-of-the-climate-apocalypse/

Noys, Benjamin. *Malign Velocities: Accelerationism and Capitalism*. Winchester Zero: Books, 2014.

Olszynko-Gryn, Jesse, and Patrick Ellis, "Malthus at the Movies: Science, Cinema, and Activism around Z.P.G. and Soylent Green." *Cinema Journal* 58, no. 1, Fall (2018): 47–69. www.ncbi.nlm.nih.gov/pmc/articles/PMC6193534/

O'Neil, Cathy. *Weapons of Math Destruction: How Big Data Increases Inequality and Threatens Democracy*. New York: Crown, 2016.

Osborne, Natalie. "For Still Possible Cities: A Politics of Failure for the Politically Depressed." *Australian Geographer* 50, no. 2, (2019).

Owens, Connor. "Social Anarchist Futures." *Solarpunk Anarchist*, May 2, 2018. https://solarpunkanarchists.com/2018/05/02/social-anarchist-futures/

Parenti, Christian. *Tropic of Chaos: Climate Change and the New Geography of Violence*. New York: Nation Books, 2011.

Pasquale, Frank. *The Black Box Society: The Secret Algorithms That Control Money and Information.* Cambridge, MA: Harvard University Press, 2016.

Passell, Peter, Marc Roberts, and Leonard Ross. "The Limits to Growth." *The New York Times*, April 2, 1972. Book Review section, 1.

Perlstein, Rick. *The Invisible Bridge: The Fall of Nixon and the Rise of Reagan*. New York: Simon & Schuster, 2014.

Polak, Fred. *The Image of the Future*. London: Elsevier, 1973.

Polanyi, Karl. *The Great Transformation*. New York: Farrar & Rinehart, 1944.

Price, Carter C., and Kathryn A. Edwards. *Trends in Income From 1975 to 2018*. Santa Monica, CA: RAND Corporation, 2020. www.rand.org/pubs/working_papers/WRA516-1.html

Quammen, David. *Spillover: Animal Infections and the Next Human Pandemic*. New York: W. W. Norton & Company, 2012.

Reich, Charles. *The Greening of America: How the Youth Revolution is Trying to Make America Livable*. New York: Penguin Random House, 1970.

Riedel, Samantha. "How V For Vendetta Predicted America's Descent into Fascism." *them.*, November 9, 2018. www.them.us/story/v-for-vendetta-fascism-america

Riesman, Abraham. "12 Monkeys is the Apocalypse Movie We Need Right Now." *Vulture*, March 13, 2020. www.vulture.com/2020/03/12-monkeys-why-terry-gilliams-movie-is-so-relevant-today.html

Robertson, James. *The Sane Alternative, A Choice of Futures*. St. Paul, MN: River Basin Publishing Co., 1978/1983. www.jamesrobertson.com/book/thesanealternative.pdf

Robinson, Tasha. "We Asked Kim Stanley Robinson: Can Science Fiction Save Us?" *Polygon*, October 20, 2020. www.polygon.com/2020/10/20/21525509/kim-stanley-robinson-interview-science-fiction-utopias

Rome, Adam. *The Genius of Earth Day: How a 1970 Teach-In Unexpectedly Made the First Green Generation*. New York: Hill and Wang, 2013.

Rushdie, Salman. "The Location of Brazil." 1985, reprinted in Rushdie, Salman. *Imaginary Homelands*. London: Granta Books, 1991.

Ryan, Mike. "'Terminator 2: Judgement Day' Is Relevant Again Today and That Sucks." *Uproxx*, January 3, 2018. https://uproxx.com/movies/terminator-2-judgement-day/

Samarajiva, Indi. "I Lived Through Collapse. America is Already There." *Gen*, September 26, 2020. https://gen.medium.com/i-lived-through-collapse-america-is-already-there-ba1e4b54c5fc

Sarkar, Saral. *The Crises of Capitalism: A Different Study of Political Economy*. Berkeley, CA: Counterpoint, 2012.

Servigne, Pablo, Raphaël Stevens, and Gauthier Chapelle, *Another End of the World is Possible: Living the Collapse (and Not Merely Surviving It)*. Cambridge: Polity, 2021.

Sherwood, S. C. M. J. Webb, J. D. Annan, K. C. Armour, P. M. Forster, J. C. Hargreaves, G. Hegerl, S. A. Klein, K. D. Marvel, E. J. Rohling, M. Watanabe, T. Andrews, P. Braconnot, C. S. Bretherton, G. L. Foster, Z. Hausfather, A. S. von der Heydt, R. Knutti, T. Mauritsen, J. R. Norris, C. Proistosescu, M. Rugenstein, G. A. Schmidt, K. B. Tokarska, and M. D. Zelinka, 'An Assessment of Earth's Climate Sensitivity Using Multiple Lines of Evidence', *Reviews of Geophysics* 58, no. 4, December (2020). https://agupubs.onlinelibrary.wiley.com/doi/full/10.1029/2019RG000678

Slaughter, Richard. *Five Steps on the Road to Recovery*, 2020. https://foresightinternational.com.au/wp-content/uploads/2020/06/Five-Steps-to-Recovery-1-2-050620.pdf

Slaughter, Richard. "Future Shock Re-assessed." *Metafuture.org*, undated, 2002. www.metafuture.org/articlesbycolleagues/RichardSlaughter/futureshock.htm

Smith, Charles Hugh. "Are You Really Middle Class?" *The American Conservative*, September 23, 2015. www.theamericanconservative.com/articles/are-you-really-middle-class/

Smith, Charles Hugh. "Our Social Depression." *Of Two Minds.com*, undated. www.oftwominds.com/blogmay15/social-depression5-15.html

Spufford, Francis. *Red Plenty*. London: Faber & Faber, 2011.

Sontag, Susan. "The Imagination of Disaster." *Commentary*, October, 1965.

Steffen, Alex. "The Apocalypse Makes Us Dumb." *Alex Steffen*, May 11, 2015. www.alexsteffen.com/apocalypse_makes_us_dumb

Sterritt, David. "Brazil: A Great Place to Visit, Wouldn't Want to Live There." Essays for the Criterion Collection, *Criterion*, December 4, 2012. www.criterion.com/current/posts/2583-brazil-a-great-place-to-visit-wouldn-t-want-to-live-there

Thompson, Edward. "Notes on Exterminism, The Last Stage of Civilization." *New Left Review* 121, May/June (1980): 3–31.

Toffler, Alvin. *Future Shock*. New York: Random House, 1970.

Turner, Fred. *From Counterculture to Cyberculture: Stewart Brand, The Whole Earth Network, and the Rise of Digital Utopianism*. Chicago: University of Chicago Press, 2006.

Turner, Graham. *A Comparison of the Limits to Growth with Thirty Years of Reality*. Canberra: CSIRO Working Paper Series, CSIRO Sustainable Ecosystems, 2008.

Tyler, Joshua. "V For Vendetta." *CinemaBlend*, May 27, 2016. www.cinemablend.com/reviews/V-Vendetta-1472.html

Urie, Rob. "Economic Decline and the Threat of Fascism." *Counterpunch*, May 1, 2020. www.counterpunch.org/2020/05/01/economic-decline-and-the-threat-of-fascism/

Villarreal, Nicolas. "How Capitalist Giants Use Socialist Cybernetic Planning." *Palladium*, September 23, 2020. https://palladiummag.com/2020/09/23/how-capitalist-giants-use-socialist-cybernetic-planning/

Wackernagel, Mathis, Niels B. Schulz, Diana Deumling, Alejandro Callejas Linares, Martin Jenkins, Valerie Kapos, Chad Monfreda, Jonathan Loh, Norman Myers, Richard Norgaard, and Jørgen Randers, 'Tracking the Ecological Overshoot of the Human Economy', *Proceedings of the National Academy of Sciences* 99, no. 14 (2002): 9266–9271. www.pnas.org/content/99/14/9266

Wallace-Wells, David. "William Gibson, The Art of Fiction No. 211." *The Paris Review* 197, summer, 2011.

Webb, Amy. *The Big Nine: How the Tech Titans and Their Thinking Machines Could Warp Humanity*. New York: PublicAffairs, 2019.

Wheeler, Issac L. "Cyberpunk is not Retrofuturism." *Neon Dystopia*, August 28, 2016. www.neondystopia.com/cyberpunk-politics-philosophy/cyberpunk-is-not-retrofuturism/

Srnicek, Nick, and Alex Williams. *Inventing the Future: Postcapitalism and a World Without Work*. London: Verso Books, 2015.

Williams, Alex, and Nick Srnicek "#ACCELERATE MANIFESTO for an Accelerationist Politics." *Critical Legal Theory*, May 14, 2013. http://criticallegalthinking.com/2013/05/14/accelerate-manifesto-for-an-accelerationist-politics/

Wright, Erik Olin. "Why Be Anticapitalist?" *Verso Books blog*, August 26, 2020. www.versobooks.com/blogs/4419-why-be-anticapitalist

Žižek, Slavoj. "Ernst Lubitsch, Censorship, and Political Correctness." *American Affairs* II, no. 3, Fall, 2018. https://americanaffairsjournal.org/2018/08/ernst-lubitsch-censorship-and-political-correctness/

Index

About the Author

Michael Harris is an experienced researcher and writer who has worked across numerous sectors including education, technology, innovation, government, public services, and urban planning. He has produced more than a hundred leading reports for organizations, both his own writing and working with some of the most important scholars across many fields. His work has been featured in national media including *BBC News*, *The Guardian*, *The Times*, *The Telegraph*, and *Prospect* magazine, as well as numerous leading academic journals.

Michael has published four books: *A Future for Planning: Taking Responsibility for Twenty First Century Challenges* (2019); *Welcome to the Rebellion: A New Hope in Radical Politics* (2020); *Stay Alive: Surviving Capitalism's Coming Hunger Games* (2021); and the current volume. He is currently based in Chiang Mai, Thailand, and works at the International College of Digital Innovation at Chiang Mai University.